T5-BBB-468

Life Cycle Assessment in Industry and Business

Springer
Berlin
Heidelberg
New York
Barcelona
Hong Kong
London
Milan
Paris
Singapore
Tokyo

Paolo Frankl Frieder Rubik

Life Cycle Assessment in Industry and Business

Adoption Patterns, Applications and Implications

With Contributions by Matteo Bartolomeo, Henrikke Baumann, Torsten Beckmann, Albert von Däniken, Fabio Leone, Ueli Meier, Renata Mirulla and Rolf Wolff

With 37 Figures and 28 Tables

Springer

Dr. Paolo Frankl
Ambiente Italia s.r.l. and
Ecobilancio s.r.l.
Via Nomentana 257
I-00161 Rome
Italy
E-mail: ecobil@cambio.it

Frieder Rubik
Institute for Ecological Economy Research
Bergstrasse 7
D-69120 Heidelberg
Germany
E-mail: frieder.rubik@heidelberg.ioew.de

ISBN 3-540-66469-6 Springer-Verlag Berlin Heidelberg New York

Library of Congress Cataloging-in-Publication Data
Life cycle assessment in industry and business: adoption patterns, applications and implications; with 28 tables/ Paolo Frankl; Frieder Rubik (ed.). With contributions by Matteo Bartolomeo ... - Berlin; Heidelberg; New York; Barcelona; Hong Kong; London; Milan; Paris; Singapore; Tokyo: Springer, 2000
ISBN 3-540-66469-6

Cover Design: de'blik, Berlin
Typesetting: Camera-ready by the authors

SPIN: 10734978 30/3136 – 5 4 3 2 1 0 – Printed on acid free paper

Acknowledgements

The study underlying this report was carried out with financial support of the European Commission (DG XII) within the European Community's "Climate and Environment"-Programme. The co-ordinator of the whole study was the Institut für ökologische Wirtschaftsforschung (IÖW) gGmbH in Heidelberg/D; further contractors were Istituto di Ricerche Ambiente Italia in Milano/I, Gothenburg Research Institute (GRI) in Gothenburg/S, Institute for Prospective Technological Studies (IPTS) of the Joint Research Centre of the European Commission in Seville/E and ökoscience Beratung AG in Zürich/CH[1]. Moreover, the study was carried out with the external collaboration of Paolo Frankl (CMER/INSEAD) in Fontainebleau/F.

The work was carried out within a project team with researchers from all the five institutes joining the project. The project was co-ordinated by Frieder Rubik (IÖW). The project team decided that two of its members, namely Paolo Frankl [CMER/INSEAD] and Frieder Rubik [IÖW], shall be authors of this book due to their final editing work. However, this work would not have been possible without the contributions of the whole team: Apart from the two already mentioned persons, it consisted of Matteo Bartolomeo [Ambiente Italia], Henrikke Baumann [GRI], Torsten Beckmann [GRI], Alexandra Bültmann [IÖW], Albert von Däniken [ökoscience], Thomas Gameson [IPTS], Fabio Leone [IPTS], Ueli Meier [ökoscience], Renata Mirulla [Ambiente Italia], Marcello Puccini [Ambiente Italia], Per Sørup [IPTS] and Rolf Wolff [GRI]. We hint at their contributions within the overview of chapters and sections (see table of contents); all chapters and sections which do not refer to specific authors are written by Paolo Frankl and Frieder Rubik.

We would like to thank everyone supporting us, especially our colleagues Susanne Nisius (IÖW), Gerd Scholl (IÖW), Cornelia Weskamp (formerly IÖW) and Matthias Saladin (formerly ökoscience). We are also grateful to our "support" staff, namely Pia Bayer, Stefan Busch, Maria Ittensohn, Alexander von Kap-herr, Patrick Lentz, Peter Naschold and Sonja Steeb; special thanks also to Jenny Bosse for her linguistic support. Also some reviewers contributed to our work and supported us by a lot of useful and interesting ideas and proposals, namely Dr. Ralf Antes (University Halle-Wittenberg/D), Prof. Dr. Robert Ayres (INSEAD Fon-

---

[1] Ökoscience got its financial support from a Swiss research programme financed by the Swiss government (Bundesamt für Bildung und Wissenschaft [BBW] and Bundesamt für Umwelt, Wald und Landschaft [BUWAL]).

tainbleau/F), Prof. Dr. Thomas Dyllick (University Sankt Gallen/CH), Dr. Paschen von Flotow (Institute for Environmental Management and Business Administration Oestrich-Winkel/D), Dr. Ruth Förster (EMPA Sankt Gallen/CH), Prof. Dr. Eckart Hildebrandt (Social Science Research Centre Berlin/D) and Prof. Dr. Rainer Züst (University Zurich/CH).

The whole research would not have been possible without the impressive willingness of 20 European companies to support us as case-study companies. The companies were ABB Italy (I), AEG-Hausgeräte (D), Akzo-Nobel Surface Chemistry (S), Bosch und Siemens Hausgeräte (D), Cartiera Favini (I), Ciba Speciality Chemicals (CH), Ericsson (S), Ernst Schweizer (CH), Fiat Auto (I), GlaxoWellcome (I), Henkel (D), Holderbank Cement und Beton [HCB] (CH), Italtel (I), Landis&Gyr Utilities (CH), Kraft-Jacobs-Suchard (CH), Norsk Hydro (S), Perstorp Flooring (S), Volvo (S), Weleda (D) and Ytong (D).

Moreover, we would like to thank Michel-Henri Cornaert from DG XII of the European Commission for his valuable remarks and support of our study.

Finally, we wish to thank two sponsoring companies, namely AEG-Hausgeräte GmbH, Nürnberg/D, and Weleda AG Heilmittelbetriebe, Schwäbisch-Gmünd/D, which supported the publication of this book financially.

Heidelberg/Rome, July 1999

Contents

1 Introduction

1.1 Life Cycle Assessment (LCA): a fascinating and sophisticated tool

The greening of the economy is not a new task, but it is a challenge for which a lot of tasks still have to be done. It is known that the main source of environmental deterioration by industry is not any more the chimneys and other process-related emissions, but the products and services produced. Products are regarded as carriers of pollution: they are not only a potential source of pollution and waste during their use; they are also a cause of resource depletion, energy consumption, and emissions during their life starting with the extraction of the raw materials and ending with their disposal (i.e. connecting production and consumption stages). The challenge of these decades is now the greening of products and services.

The new focus on products (cp. Oosterhuis/Rubik/Scholl 1996) was introduced as a policy approach of shared responsibility in which different actors are involved along the life-cycle of a product, each having specific responsibilities. Their common objective includes the reduction of the environmental impact of products. The 5th Environmental Action Programme of the European Union (EAP 1992) included several policy guidelines, which embrace

- *orientation towards causes* (environmental policy should focus on the agents and activities which harm the environment rather than confine itself to reactive repair of environmental damage),
- *shared responsibility* (by all involved actors, including public administration, public and private enterprises and the general public),
- *a broadening of the range of instruments and their combination* (legislative, market-based instruments, horizontal, support instruments and financial support mechanisms).

The move is away from the command-and-control policy to the pluralistic approach of shared responsibility among different actors. Thus enterprises are given much more responsibility and freedom with the objective to stimulate proactive and preventive behaviour of business as partners for “sustainable development”. The government’s task is restricted to providing the framework of environmental policy and to encourage the greening of product management by business.

Different tools, concepts and ideas to support this have been developed in the last 20-30 years. In this report, we examine one concept, namely **L**ife **C**ycle **A**s-sessment (LCA), which is perceived as a fascinating but also sophisticated tool for a greening of business and economy. LCA is among the most instructive management tools for getting insights into product related environmental impacts. It is also compatible with the concept of extended producer responsibility.

LCA is a tool for assessing the potential environmental impacts of products[1]. It assesses products from several perspectives:

- the whole life cycle of the product, i.e. combining an upstream with a downstream focus,
- consumption of different input streams, e.g. non-renewable resources and
- cross-media outputs, i.e. considering environmental burdens in several ecological media.

Its history is a relatively new one: beginning in the late 1960s, its real breakthrough into the business world occurred in this decade. Enthusiasm and scepticism alternates: LCA has been hailed as the "environmental management tool of the 1990s", "LCA will be seen as an integral part of the environmental management tool-kit" (Jensen et al. 1997, p. 28). LCA procedures are too expensive and complicated; they could only seldom be used (Arnold 1993). Contrary to that, Berkhout (1996) mentions: "Adoption of life cycle approaches is a sign of beyond compliance behaviour" (p. 151). There is a lot of controversy in the scientific discussion. Nevertheless, the progress on the level of the international standardisation at ISO is impressive: two standards have been agreed upon, several others are under way[2].

The ISO-standard 14040 defines an LCA as following: "LCA is a technique for assessing the environmental aspects and potential impacts associated with a product, by:

- compiling an inventory of relevant inputs and outputs of a product system;
- evaluating the potential environmental impacts associated with those inputs and outputs;
- interpreting the results of the inventory analysis and impact assessment phases in relation to the objectives of the study" (ISO 1996, p. iii).

According to this standard, an LCA consists of four different steps:

1. Goal and scope definition,
2. inventory analysis,
3. impact assessment,
4. interpretation.

Applicants of LCA are a plethora of different actors and groups, such as business, business associations, government, public agencies, NGOs and researchers.

---

[1] The term "product" should be interpreted in a broad sense and includes also services.

[2] ISO-standards 14040 and 14041. The standards on impact assessment (14042) and interpretation (14043) are being worked on and might be passed during 1999 or 2000 (see also Box C in Chapter 7).

LCA can be used for quite different purposes: for bottleneck identification and product innovation in industry, for marketing and information of consumers, for strategic planning within companies and for policy making.

In practice, the application of LCA through policy has lead to controversy among the different stakeholders, companies and the state. There are considerably different views on it. German experiences of LCA on packaging show that there is a lot mistrust and fear with respect to the application of LCA by political actors and public institutions.

In fact, companies are the main subjects which determine both the economic and the environmental life-cycle of products[3]. Once a product has been designed and produced, both the economic and the environmental life-cycle have already been very much determined. Of course, they are influenced by the behaviour of users and of the people responsible for waste management. But these are mainly determined by the early steps within an enterprise. The crucial factors are the design and product development phases. Pfeiffer (1983, p. 67) argues using the example of mechanical engineering that more than 70% of the production costs are determined by product development. The US National Research Council (1991) estimated that more than 70% of the costs of product development, its production and its use is determined by the design steps of a product. In the same way quite a lot of the environmental burdens of a product along its environmental life-cycle are pre-determined during the design & development phase. This means that companies have to consider as early as possible environmental aspects, in order to improve the eco-profiles of their products.

LCA supports business in their consideration of environmental issues. It helps them to delineate their share of product responsibility and the results bring the environmental dimension into the business decision making process. However, companies remain confronted with complex decision-making situations integrating many aspects, such as technical and economic matters, risks, work environment or safety issues.

The research on LCA during the last years was and is still focused on improving the methodology and integrating diverging interests. Indeed, the "LCA-technique" has improved considerably. Especially the Society of Environmental Toxicology and Chemistry [SETAC] has pushed this process by preparing a "Code of Practice" (SETAC 1993) and by organising a series of international workshops in Europe and the United States. Since 1995, the "International Journal of LCA", an international research journal which deals exclusively with LCA,

---

[3] The *economic* life-cycle is described in several ways. The demand life-cycle encompasses emergence, accelerating growth, decelerating growth, maturity and decline (cp. Kotler 1991, p. 348f.).

The *environmental* life-cycle considers a product from cradle to grave, i.e. it includes the extraction of the raw materials, the different production and distribution processes, the use/consumption of the product, until the latter becomes waste and will be treated for reuse, recycling, recovery, disposal or land filling.

exists. These activities, a lot of workshops, conferences, meetings and research activities, have contributed to the elaboration of international standards.

However, the very important methodological discussions and progress of the last 5-10 years have concentrated on the improvement of the method as such. The embeddment of the method into the application context has not been stressed nearly enough; this is a new challenge. New open questions arise: What is the contribution of LCA to decision-making processes? How does business deal with LCA results? Which departments of a company use LCA? Do they have specific demands? What is the relationship between the environmental dimension represented by LCA and other dimensions? Who promotes LCA within companies? A lot of questions which might easily be extended.

Some research has been undertaken just recently. Rubik/Grotz/Scholl (1996) explored potential environmental benefits of LCA; also the demand for "streamlined" techniques have been supported by several publications (e.g. SETAC 1997). Nevertheless, the questions with regard to the role of LCAs within the decision-making processes have hardly been examined. First experiences have been delivered by FTU/VITO/IÖW (1995a and b) and Baumann (1998). The objective of this report is therefore to improve the state-of-the art research in this direction. *The primary focus of this report is dedicated towards the application of LCA within business and the ways that LCA is used.*

1.2 Objectives and scope of this book

The report presented herewith aims to examine the application of the instrument of Life Cycle Assessment (LCA) within business and the implications of its use for environmental policy. The following main research questions are being addressed:

1. How is LCA used within business decision-making processes?
2. Which other tools influence these processes?
3. How can different decision-makers consider the environmental dimension in decision making while confronting other aspects?
4. How does LCA influence the distribution of environmental knowledge and environmental information?
5. How does LCA fit with different environmental management structures in the context of decision making?
6. Does LCA require other instruments or methodologies to be an effective decision-making support? (barriers and opportunities)
7. Are there different application patterns within the European Union?
8. How can the use of LCA be improved considering existing barriers and opportunities?
9. What are the expectations of business *on* policy makers?

Altogether we focus on two key issues:

- On the one hand, we concentrate on the use of LCAs within business decision-making processes. The influence of LCA on business decision-making processes is analysed within the framework of this project.
- On the other hand, we examine the interactive relevance of LCA to politics, i.e. the expectations of business regarding policy-making activities and of policy-makers regarding the business use of LCA.

The main focus of the report is dedicated to the first issue.

Methodologically, the study uses a number of approaches. At first, a theoretical framework for decision-making has been developed; this framework presents the state of the art. Following this, two different empirical approaches on the ways business deals with LCA have been carried out:

- A survey which was carried out in spring 1997 based on a questionnaire that was standardised and used for the four selected countries (Germany, Italy, Sweden and Switzerland).
- Through a set of at least 20 case-studies into the use of LCA in business (at least five case-studies for each of the selected countries: Germany, Italy, Sweden, Switzerland). These case-studies include enterprises of different branches and sizes.

One can compare some important characteristic aspects of both methods, see Table 1.1.

Table 1.1. Characteristic factors of methods used

Aspect	Survey	Case-studies
Addressed companies:	1 625 companies	40 companies
Received positive answers:	382 companies	20 companies
Approach:	Quantitative approach	Qualitative approach
Sources:	Single source: • Written questionnaire	Multiple sources: • Documentation • Analysis of literature • Personnel interviews • Observations • Use of background information
Interviewees per company:	One employee answering the questionnaire	Several employees
Main characteristics:	• Enlarged examination • Static analysis • No further explanations possible • Consideration of aspects which are appropriate to be asked in a written questionnaire. No supplementary information possible	• In-depth examination • Dynamic analysis • Questions, explanations, further questions possible • Consideration of details, hidden facts, ideas and interesting aspects found during research process

Of course, the empirical bases of both methods are complete different. A survey is able to consider a much larger number of companies whereas a case-study analysis has to be restricted to a very selected number of cases. On the other hand, the advantages of case-study research are that they deliver a clearer and more comprehensive picture of the examined companies (see Yin 1994).

Both empirical research methods deliver complementary results and enlarge our empirical findings.

Beside the examination of the application of LCA within business, we have explored the expectations on the use of LCA among business and policy by analysing the situation in the four countries represented in the group, namely Germany, Italy, Sweden and Switzerland. To do this, policy makers were interviewed.

The whole report is not primarily intended to improve the method of LCA. It intends to look at the integration and implementation of LCA within business; it regards the tool as externally given and considers the institutionalisation of an innovative approach like LCA within the business world. Our basic approach is to explore and improve the application-context of LCA. Therefore, our overall approach comes from the social sciences and not from the natural sciences.

1.3 Outline of this book

This volume is subdivided into six chapters. It is structured as follows:

Chapter 2 Framework and Theoretical Background develops a conceptual framework to understand and study LCAs in relationship to management processes, i.e. to study the use of LCAs as a tool in business; chapter 2 highlights the main aspects of the use of LCAs in a business context. Three questions are answered, namely how is LCA related to decision making, how is LCA related to management processes and what are the conclusions for the empirical study of the use of LCAs? The second part of the second chapter presents the tool LCA as such and illustrates the application possibilities of LCA in business; to do this, we present the applications according to the product development chain and to the direction of efforts (internal vs. external).

Chapter 3 Application of LCA in General presents an overview on the application of LCA by different users and actors in four different European countries, namely Germany, Italy, Sweden and Switzerland; the state of this information is as of the end of 1996.

Chapter 4 A "Static" Perspective on LCA Applications – Survey Results summarises the results of the first empirical approaches performed, namely a survey with a questionnaire that was standardised and used for the four countries mentioned above (Germany, Italy, Sweden and Switzerland).

Chapter 5 The Dynamics of LCA Adoption and Integration in the Firm – The Results of the Case-Studies summarises the results of our set of 20 case-studies which explored in detail the use of LCA in business in the four selected countries Germany, Italy, Sweden, Switzerland. These case-studies include enterprises of different sectors and sizes. For both empirical parts, we introduce preliminary conclusions in this chapter 5.

Chapter 6 The Relationship between Business and Policy: Expectations and Implications is restricted to one specific and important aspect of the role of policy in LCA, namely towards the expectations between business and political actors on the application of LCA by business. We look both at the business expectations regarding policy-making activities in the area of LCA, and at the policy level and their views on the business application of LCA.

Chapter 7 General Conclusions and Recommendations reviews the results of the study and draws a number of conclusions and recommendations. The conclusions are subdivided into a summarising presentation of the results of the whole study and into a presentation of the main findings. The recommendations have been subdivided into those mainly directed towards environmental policy makers at the various levels, and those primarily addressing further research needs.

2 Framework and theoretical background

This chapter develops a conceptual framework to understand and study LCAs in relationship to management processes, i.e. to study the use of LCAs as a tool in business; first, we highlight the main aspects when it comes to the use of LCAs in a business context. Questions on the relation of LCA to decision making and to management processes are analysed and some hypotheses for the empirical part of work are developed (section 2.1). The next part of this chapter (section 2.2) presents the tool LCA as such and illustrates the application possibilities of LCA in business; to do this, we present the applications according to the product development chain and to the direction of efforts.

2.1 LCAs as reflective conversations with messy situations: a decision making appraisal

2.1.1 Introduction

One of the aims of LCA is to support decisions concerning operations, products and strategies in business. Much attention and effort has been given to developing the tool as such, but knowledge about how the use of LCA affects management is missing.

In this, we develop a conceptual framework to understand and study LCAs in relationship to management processes, i.e. to study the use of LCAs as a tool in business. Although we cannot develop this to the extent we would like to highlight the main aspects when it comes to the use of LCAs in a business context. To do this, we have to answer three questions:

1. How is LCA related to decision making?
2. How is LCA related to management processes?
3. What are the conclusions for the empirical study of the use of LCAs?

In the following part, we discuss the relationships between decision making, information processing, managerial work and LCA.

2.1.2 Decision making and information processing: a short review

Since the work of Simon (1947), the limits of human rationality have been viewed as institutionalised in the structure and modes of the functioning of organisations. Individuals and thus organisations have limited information-processing abilities, because of

- incomplete information about possible courses of action,
- their ability to only explore a limited number of alternatives relating to a given decision and
- their inability to attach accurate values to outcomes.

"Bounded rationality" and "good enough decisions" are based on simple rules of thumb and limited search and information.

It is also acceptable in organisation and management theory to differentiate between operative and strategic decisions. Already in 1958, March and Simon made the important distinction between standard operating procedures (SOPs) and decisions that are unique and most probably creative in character (March/Simon 1958). Whereas SOPs are based on an organisation's memory, strategic decisions have to develop new answers to unique situations.

This distinction has endured some decades, but now there is a growing awareness amongst management researchers that this distinction may not reflect the actual structure of decision making in organisations. Organisations are now less hierarchical in structure and one separates operational from strategic decisions mostly for analytical reasons; in reality these are more and more interrelated in day-to-day actions. Since organisations are now more knowledge-intensive (Nonaka/Takeuchi 1995), less hierarchical and rather boundaryless (Hirschhorn/Gilmore 1992), important decisions are taken and/or performed at all levels of the organisation. Yet, the power to discriminate strategic from non-strategic lies in the hands of a limited number of people in the organisation and thus the centre has greater influence upon the outcomes of and the choices as such.

It is reasonable to assume that LCAs as learning instrument are at present mainly concerned with decisions that are uncertain, contain elements of learning and possibly change within organisations. Thus, LCA related decisions are not SOPs. Consequently, when looking at LCA as a decision tool and focusing upon its use, rather than its validity as a tool, these aspects have to be taken into account. Questions for research related to these assumptions are for example: Who initiates the conducting and implementation of an LCA? For what purpose? In what context? With what results for processes or products?

2.1.2.1 Rationality and choice

Decision making has been described in many ways in literature. One of the predominant schools of thought in management sciences is led by the assumption that decision making amongst managers is rational or quasi rational. Rationality is viewed as the relationship between given ends and the means chosen to achieve these ends. It is implied in this view, that managers choose one best alternative, based on the evaluation of several alternatives at hand. This rather basic structure has pre-occupied management sciences whose proponents developed operation research as the scientific support for rational decisions. Organisation theorists on the other hand focus on understanding the empirical behaviour of decision makers rather than on developing mathematical models. The rational tradition often follows a normative path, whereas organisation theory usually employs a descriptive approach to increase our understanding of everyday life processes.

Three schools of thought regarding strategic decision making in business can be distinguished:

- rational normative school,
- strategic choice,
- external control.

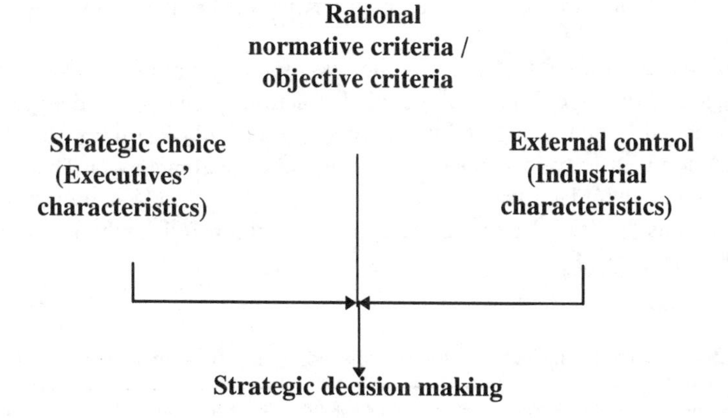

Fig. 2.1. Three modes of strategic decision-making (simplification of Hitt/Tyler 1991, p. 328)

Special attention is given in this context to the institutionalisation processes within the external control perspective. How are norms and methods of LCA spread in industry? Before the use of LCA becomes fully institutionalised, i.e. taken for granted and routine, a process which goes through certain stages has to take place. Tolbert/Zucker (1996) describe these stages in their comprehensive review:

- innovation,
- habitualisation,
- objectification and
- sedimentation.

Habitualisation involves the development of patterned problem-solving behaviours, and the identification of signals for when to apply the problem-solving behaviour, i.e. by using LCA. At this stage, technical and economic factors together with internal political arrangements largely predict adoption. On the whole, there are generally few adopters at this stage. They can be found in interconnected organisations facing similar circumstances. However the knowledge among non-adopters about LCA activities is extremely limited.

Objectification is probably the most crucial stage in the institutionalisation process. It involves the development of a consensus regarding the problem-solving behaviours, here the use of LCA. The two mechanisms from which consensus can emerge are:

- through information gathered from many sources in order to confirm adoption, and
- through a champion who has to provide successful cases and to justify the use of LCA as a solution to a diagnosed organisational problem.

Technical/economic/political characteristics are no longer as important, and activities may follow the fashion-like quality. Adopters are therefore more heterogeneous.

Sedimentation is characterised by the cessation of promoting and accumulating evidence and by the existence of a formal structure spread across the group of actors theorised to be the appropriate adopters. When fully institutionalised, the use of LCA survives across the generation of organisational members. This is because new members who lack knowledge about the origins of LCA activities are apt to treat them as "given". Factors important for reaching full institutionalisation are described in Table 2.1.

Table 2.1. Factors that influence institutionalisation (adopted from Tolbert/Zucker 1996)

1. The larger and more centrally linked the organisations which are innovators and early adopters are, the more likely it is that the structure will become institutionalised.
2. As the status of those in opposition increases, the degree of institutionalisation decreases.
3. The broader the range of applications for which a given structure is theorised to be relevant, the more difficult it should be to provide convincing evidence of a structure's effectiveness, and hence there will be a lower level of institutionalisation.
4. The greater the number of champions, the less likely it will be for entropic processes to become operative, and thus there will be a higher level of institutionalisation.
5. Higher investment costs should also mitigate entropic tendencies, thus resulting in a higher degree of institutionalisation.

Implications for LCA

Given the rational model, it could be assumed that LCA will sometimes be used as a rational analytical tool to improve decisions regarding the ecological impact of a company's activities (operations, products, strategies).

The strategic choice model suggests that ecological considerations depend much more on the attitudes and personal dispositions of the top management of a company.

The external control perspective argues that a company's consideration of ecological facts relates much more to the industry's field and thus to the competition the company is involved in. Special attention is given in this context to how norms and methods of LCA are spread.

2.1.2.2 Talk, decision and action – focus on learning

The environmental debate has been overloaded with myths and accusations. Government accuses business of failing to comply with emission levels; business accuses government of its lack of understanding of business conditions; consumers are accused of not purchasing ecologically adapted products.

The misunderstandings underlying these conflicts are to some extent based on ideologically-rooted judgements but they also stem from a lack of understanding of the contingencies surrounding every actor system. It is always assumed that actors act either irrationally or hypocritically (Brunsson 1994). One way of clarifying this messy picture of accusations is to increase our understanding of the action alternatives which every system possesses: talk, decision and action (March 1986; Brunsson/Olsen 1993; Czarniawska/Wolff 1986). LCA as a learning tool may provide processes that enable companies to gain time to learn more about their impact on the environment.

Table 2.2 demonstrates how these three alternatives for dealing with environmental demands can be exploited by various actors to create learning opportunities in each system. Talk can be used by the political system to create awareness in firms, for example, that a particular environmental issue has to be tackled. Talking can be used even when there is no certainty about what the necessary solution or decision will be; talking can also prevent controllings agencies and instruments from becoming institutionalised.

Table 2.2. Learning arenas and how actors learn (Andersson/Wolff 1996)

Learning arenas: / Parties involved:	**Talk**	**Decision**	**Action**
Government	Environmental debate to push and learn	Laws and taxation	Institutionalised control
Industry	Public affairs (legislation & decoupling)	From policy to strategy	Process and product development
Consumers	Intentions	Routines (such as shop loyalty)	Using new infrastructure

Uncertainty is a crucial element related to information processing. Organisations tend to respond to uncertainty in two ways (Galbraith 1973), either through the creation of slack (and time buffers are one way of slack creation) or by investing in sophisticated information systems (and LCAs could be one way of creating complex information about process and product environmental impact).

Consumers can reduce uncertainty by letting retailers decide on selecting ecological products and discriminating against non-ecological items. By decoupling learning arenas (talk, decision and action), actors can create learning opportunities and space for developing environmental solutions, as well as legitimising their actual environmental problems. It is one of the important tasks of research to contribute to an understanding of these mechanisms and the trade-offs between them.

Implications for LCA

By relating LCA activities to these three arenas, we can ask businesses which actors use LCAs and for what purposes. Is LCA used by environmental managers as a way of forcing top managers to change strategies for products? Is LCA used to learn, i.e. to mirror emissions and pollution? Is LCA simply used as a legitimising device?

2.1.2.3 Research on the nature of managerial work and implications

In the area of "environmental management", there is a pre-dominance of model and tool development, rather than a focus on the *use* of tools in everyday management practice. From the research on managerial work, insights can be used to 'place' LCA models in management processes.

Elsewhere (Wolff 1997), a framework for the sensible use of environmental tools has been developed which considers their intended purposes. This framework has been adapted to the factual situation most managers face. Studies on managerial work tend to look at the substantive elements of managerial work, the distribution of managers' time between different elements, whom managers interact with, informal aspects of managerial work and important themes that managers work with (Hales 1986, p. 90). Most studies seem to conclude that much time is spent on day-to-day trouble-shooting and ad hoc problems of organising and regulation. On a general level, studies on managerial work claim that:

- managerial work is fragmented,
- managers have a lot of different roles to play,
- managers deal with many conflicts,
- managers take decisions under severe time pressure,
- strategic decisions tend more and more to be taken intuitively,
- and thus are less based on intensive analytical analyses.

Given these preconditions, we have to ask what role LCAs can play to support decision makers when managerial work is fragmented and constantly disrupted. This remains a challenge for empirical research.

2.1.3 LCA in business practice – some first empirical research results

The use of LCA in industry has been studied through questionnaire surveys such as the Business Environmental Barometer of 1993 and 1995, as well as the one in this project. The results from the surveys indicate a rather broad use of LCA (Baumann 1996), see Table 2.3. For the results of the survey presented within this report, see chapter 4.

Table 2.3. Areas of applications for LCA in listed Swedish companies in 1993 and 1995

Areas of applications	1993	1995
• Analysis of own product	12	14
• To learn about LCA	8	n.d.
• Product development	6	11
• For external use (marketing, labelling...)	6	11
• Process development and optimisation	5	9
• Choice of suppliers and raw materials	5	9
• In training programmes	5	5
• Analysis of line of business	3	8
• To meet authorities' demands	3	6

n.d. = not defined

There seem to be few examples where LCAs support a single decision. Instead, LCAs seem to be used in order to understand complex situations, which is referred to as "learning" in organisational literature. Such use of LCAs is also mentioned by Bakker (1995), who found that to suit the informational needs of industrial designers, LCAs should not be too narrow in scope. Preferably, LCAs should be on a product system level rather than on a product level so that the LCAs provide examples for learning rather than support for specific decisions. In organisational literature, this is sometimes referred to as "double-loop learning" (Argyris/Schön 1976). With this type of learning, new knowledge complements or alters the frame of reference of the designer. As the new knowledge becomes integrated in the designer's "bank" of experience, it is applied in subsequent design projects whenever the designer draws conclusions based on prior experience. The same situation applies to managers in industry faced with time pressure and fragmentation of tasks (Mintzberg 1973; Jönsson 1995). The way in which LCA can support decision makers in these situations is by providing learning examples and by helping with the structuring of problems and situations, rather than being used as a decision tool.

Discussing theories of decision making and organisational choice, March (1981) concludes: "The intelligence of organisational action is seen as lying not in the capability of knowing everything in advance but in the ability to make marginal improvements by monitoring problems and searching for solutions. Thus, theories of limited rationality are essentially theories of search and attention: what alternatives are considered? What information is used?" (March 1981, p.5).

In this context of limited rationality, as also indicated in the survey results, LCAs provide important search methods and add information to the decision process in organisations.

2.1.4 The conceptual framework: environmental management as the management of messes

2.1.4.1 Three types of judgement in messy situations

Environmental management, i.e. the management of environmental concerns in a company, has yet to be defined. Whereas environmental demands now have a long history of research and debate, very little has been said about the management aspects of them. Ackhoff (1979), in a critical discussion of operation research and the management sciences in the US, has characterised management as the management of messes: "Managers are not confronted with problems that are independent of each other, but with dynamic situations that consist of complex systems of changing problems that interact with each other. I call such situations *messes*. Problems are abstractions extracted from messes by analysis; they are to messes as atoms are to tables and chairs. (...) Managers do not solve problems: they manage messes" (Ackhoff 1979, p. 99f.).

"Messes" are further described as "systems of problems". As the sum of the optimal solution to each component (= problem) taken separately is not an optimal solution to a mess, the internal dynamics between the problems are much more important for solutions, than the solution of problems independent of each other.

Schön (1983, p. 14) describes practical situations with some basic features: they are complex, uncertain, unstable, unique and they contain value conflicts. Practical situations are therefore often "messes" with certain fundamental features; they are not problems and as a rule there are no simple solutions to them. One could say that managing is dealing with a constant and ever-growing flow of messy situations.

The actual debate in "environmental management" focuses mostly on "solutions", i.e. the development of both simple and rather sophisticated tools. There are in principle two types of "solutions" (Wolff 1996):

- "ecological solutions": these assume that the environment demands certain actions and that these actions have to be enforced on business. If business does not respond to these ecological demands, this has to be enforced on business by means of legislation or other forces.
- the "tool solutions": it is implied that "good" tools are "solutions for a better environment". Hence, these tools have to be implemented.

Of course, there is a need for good tools, but the question is how sophisticated should these tools be, given the presence of complexity, uncertainty, uniqueness, value conflicts and instability that characterise most decision situations?

Vickers' work on decision making (1965/1995) enables us to extend the framework regarding organisational choices. Based on his research, he developed the notion of "appreciative" systems. Such systems contain three types of judgements:

- reality judgements,
- value judgements, and
- instrumental judgements.

Reality judgements answer the question of "what is", value judgements of "what ought to be" and instrumental judgements try to solve the mismatch between the first and the second. In order to systematise learning related to each type of judgement, processes of prediction, change and innovation and evaluation have to be implemented.

It seems to us that LCAs do have qualities to support reality and instrumental judgements since LCAs describe process or products impact on the environment. By doing so reality descriptions are gained. LCA also enables us to make sensible judgements regarding choices between alternatives, for example between materials.

2.1.4.2 LCA as a tool of judgement in messy situations

Managers are probably less concerned with the solution, but more with the structuring of problems and situations. This may be the foremost impact LCA can make in business. Schön claims that decision makers are concerned with "design" as in "design as a reflective conversation with the situation" (1983, pp. 76-104). He builds on Simon's notion of the structuring of complexity (1977); all professionals who would like to accomplish transformations of existing situations to the better are occupied with "design". Designing the interactions between the problems that constitute messes and making sense out of these is an important management task, which Vickers labels as "judgements".

The LCA debate has shown that LCAs can take many forms. On the one hand, LCA is a "holistic" concept, on the other a highly detailed numerical model developed in an LCA study/project. There are a range of more or less detailed LCAs, as is shown by the following set of terms used in the debate (see also section 2.2):

- quantitative LCA (full LCAs or simplified/streamlined/screening LCAs), with a computational algorithm;
- qualitative (descriptive) LCA, with e.g. a matrix methodology;
- conceptual LCA (Life Cycle Thinking [LCT]).

This range of methodologies can be thought of as reflecting different ways of structuring "messy" situations.

Simon et al. (1987) describe decision making as consisting of processes for problem solving, the assessment of alternatives and learning. The way in which a decision is organised and carried out depends, therefore, a great deal on the experience of the decision maker. The three judgement processes can be more or less present in different LCA applications. The LCA methodology in these different LCA applications also reflects these differences. For instance, process development, which is mainly characterised by the problem solving process, is perhaps best served by a stand-alone LCA used in an exploratory way to identify and generate alternatives. In the purchasing situation, on the other hand, when the assessment of alternatives can be the dominating process, comparative LCAs may be used.

The influence of the learning process is discussed by Argyris/Schön (1976) in their model of single-loop and double-loop learning. Single-loop learning includes problem solving and the assessment of alternatives, while double-loop learning also includes transformational learning. The former implies an application of knowledge and models, whereas the latter implies a reflective decision maker who through learning may redefine his/her framework, and allow for an "unorthodox" application of models.

Thus, for LCAs to be used intelligently as a design of situations, rather than as a decision tool, LCA methodology will need the supplementary knowledge of decision behaviour, such as was touched above.

Given that LCAs can function in the messy situations of judgements, the research challenge on the use of LCA is related to the empirical data on the nature of managerial work. Given the fact of time pressure and fragmentation, is decision making supported by LCAs? Furthermore, how can decision making be supported by LCA-results?

2.1.4.3 Conclusions

This very short overview of some important aspects of decision making and managerial work gives us some clues about the empirical challenge related to study the use of LCA in business. It serves as a guide for focusing, not on the rational technical aspects (the "shoulds"), but on the actual use and functions LCAs play when decision makers invest in conducting them.

Table 2.4 summarises how to empirically study the use of LCA. It is based on the acceptance that LCAs do play different roles, from rational decision support to symbolic ones (when conducting an LCA is an act of cultural significance). With this in mind, we have to differentiate between the rational use of LCA models and the more judgement-based use of the LCA model as described above. In addition, we stress the importance of contextual aspects related to national, industrial and company characteristics.

Table 2.4. Areas of research and research strategies

Mode	*Contexts*		
	Nation	**Industry**	**Company**
Rational problem solving: • products • processes	Study of LCA diffusion	Survey studies of industry	Case study research
Judgements in messes: • reality judgements • value judgements • instrumental judgement	**Participative process studies** ("ethnographic" studies)		

In this publication, we present the results of research conducted by means of a survey to describe the general use of LCAs (in chapter 4) together with those of twenty exploratory case studies made to gain a deeper understanding of the industry-related use of LCA (in chapter 5).

2.2 LCA Methodology and its main applications

2.2.1 Introduction

The aim of this chapter is to describe Life Cycle Assessment and its main applications. The introductory subsection 2.2.1 focuses on the concept of LCA. Very often, people use this acronym for very different things, ranging from very vague approaches based on life-cycle thinking, to full LCAs as codified by ISO standard 14040. Subsection 2.2.1.1 clarifies the differences between the concepts of Life Cycle Assessment, various Life Cycle Approaches and simple Life Cycle Thinking. Subsection 2.2.1.2 focuses on the tool LCA itself (definitions, phases of an LCA, key features and limitations, screening and/or streamlined LCA) as described and codified by ISO standard series 14040. The second subsection 2.2.2 tries to list and give examples for the main applications of LCA. The classification is made according to the product development chain: subsections 2.2.2.1 to 2.2.2.4 focus respectively on the applications of LCA for strategy definition, research, development and design, procurement and production, and marketing. Finally, subsection 2.2.2.5 discusses the use of LCA for information, training and education purposes.

2.2.1.1 Life Cycle Assessment and Life Cycle Approaches

In order to implement an effective and efficient product environmental innovation policy, a company has to consider the whole life cycle of a product. Instruments to be used by companies to enhance an effective product innovation policy can differ widely from company to company. When using the acronym LCA researchers usually refer to Life Cycle Assessment, a sophisticated environmental impact assessment tool that comprises a number of steps (see next subsection 2.2.1.2).

Companies' experience suggests that LCAs are costly and time consuming because they are inherently complex and data intensive (Sustainability et al. 1993), are subject to technological change and are dependent on data which is often difficult to access. This is the reason why, apart from large corporations, businesses frequently prefer to use more simple instruments based on the same life cycle approach or Life Cycle Thinking, but entailing lower efforts in terms of data acquisition and analysis, impact evaluation [1].

Basically, there are three different levels at which the concept of a product's life cycle can be taken into account (see Figure 2.2).

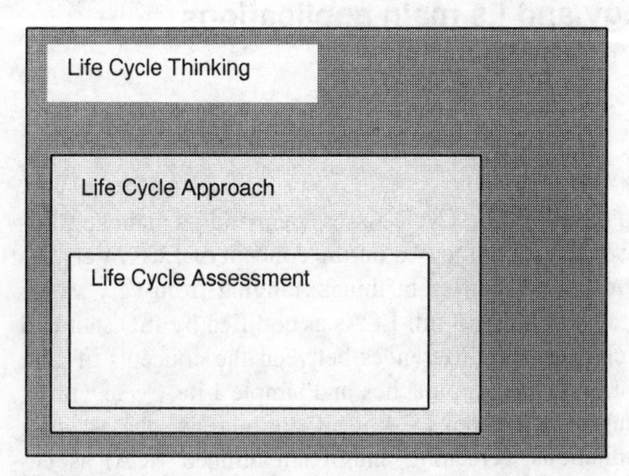

Fig. 2.2. Basic levels of assessing the life cycle of a product

First, there is the broad framework of **Life Cycle Thinking** (LCT). At this level, the concept of the life cycle is used as a rather general and rather qualitative approach within a broad framework encouraging and promoting a proactive behaviour of business towards the environment. The essence of LCT is that products and processes do not exist in isolation, but carry baggage and implications. When co (type 3)nsidering products, one cannot ignore either preceding – upstream –

[1] This section 2.2 focuses on both LCA and life cycle approaches; the acronym LCA will be used with both meanings (i.e. Life Cycle Assessment and Life Cycle Approaches).

activities (extraction, processing) or downstream activities (dismantlement). Among others, this has been very clearly stated by Schmidt-Bleek et al. using the rather evocative term "Rucksack" (Schmidt-Bleek et al. 1994). However, within LCT, the life-cycle environmental impacts and implications are treated only in a qualitative or simplified way.

The level of Life Cycle Approach is the first one at which companies operationalise this thinking.

Life Cycle Approaches include a wide range of instruments (OECD 1995); a list of some major instruments based on life cycle approach is given below:

- LCI – **L**ife **C**ycle **I**nventory – is a single step of LCA itself and includes the identification of physical flows related to the different phases of the product life cycle (see also subsection 2.2.1.2.2). Properly combined with improvement analysis it can provide managers with useful information for bottleneck identification, material and energy stream reduction, and other product environmental improvements. However, LCI alone cannot provide information on actual environmental impacts. Particularly in the past, many LCAs carried out by companies were actually limited to the LCI phase only.
- SLCA – **S**treamlined **L**ife **C**ycle **A**ssessment – and BLCA – **B**ottleneck LCA – are simplified LCAs (VROM 1994; SETAC 1997) where main flows between the product (along its life cycle) and the environment are defined in a simplified and sometimes only qualitative form. This application of the concept of life cycle evaluation has the aim of providing decision makers with very basic and cheap information on main product impact on the environment. SLCA is frequently performed when it is considered that the relevant data to perform a real LCA will not be available. SLCA (VROM 1994) has also been used by the Dutch government as a basis to define ecological criteria under the ecolabel scheme[2].
- LCC – **L**ife **C**ycle **C**osting – consists of an inventory and analysis of economic implication of environmental impact of a given product during its life cycle. The term Life Cycle Costing has different meanings and can include the identification of internal costs (for a company) and/or external/social costs.
- MIPS (**M**aterial **i**ntensity **p**er **s**ervice unit) (see Schmidt-Bleek 1994) and eco-point evaluations (see Ahbe/Braunschweig/Müller-Wenk 1990) are methods presenting a single aggregated number to describe the use of material and the impacts of the product. The first approach has been developed by Wuppertal Institute in Germany, the second one by Federal Office for Environment, Forest and Landscape [BUWAL] in Switzerland. The method aggregates the impact parameters like air pollutants, water pollutants, energy consumption and waste production to one single figure (MIPS for inputs only). Even if a single figure is considered as very attractive by some researchers, practitioners and consumers, difficulties in the implementation have largely reduced the use of these systems.

---

[2] For a more detailed discussion of streamlined LCA refer to subsection 2.2.1.2.4.

- Eco-labelling criteria are based on Life Cycle Approaches and therefore focus on the optimisation of a few , relevant material flows related to product life cycle. This narrow scope definition has the aim of making the eco-labelling schemes viable in terms of criteria definition and company compliance.
- Finally, many conventional analytical tools in companies, which were originally developed as supporting tools for decision-making long before the upsurge of environmental issues, are now developing along the Life Cycle Approach (OECD 1995). Some major tools are:
  - cost-benefit analysis,
  - technology assessment,
  - multi-criteria analysis,
  - environmental accounting,
  - environmental auditing,
  - environmental risk analysis and assessment,
  - hazard assessment,
  - energy and material analysis,
  - environmental impact assessment,
  - waste minimisation.

All these approaches reflect the increasing consciousness in companies of the complexity of product and process interactions with the environment (OECD 1995).

Finally, there is the level of **L**ife **C**ycle **A**ssessment (LCA) itself. This rather well-coded tool is described in the next subsection.

2.2.1.2 What is Life Cycle Assessment?

2.2.1.2.1 Definitions

The procedures for initiating, conducting and reporting LCA studies in a proper manner have been defined by several international organisations during recent years. Many workshops have been carried out on LCA since 1990. In particular in 1993, in response to an increasing need for guidance in LCA, the European and North American organisations of the Society for Environmental Toxicology and Chemistry (SETAC) organised a "Code of Practice" Workshop in Portugal. The outcomes of the workshop were summarised in a booklet called "Guidelines for Life-Cycle Assessment: A Code of Practice" (SETAC 1993).

More recently, the guidelines and principles relating to LCA studies were defined by ISO/TC 207/SC 5 working group using specific international standards, namely ISO 14040 (ISO 1996). Methodological details are reported in the supplementary ISO standard 14041 which has already been accepted and the Draft International Standards (DIS) 14042 (Impact Assessment) and 14043 (Interpretation)[3].

---

[3] See also Box C for an actual overview on the state of the ISO 14040-series.

The ISO-standard 14040 defines an LCA as following: "LCA is a technique for assessing the environmental aspects and potential impacts associated with a product, by:

- compiling an inventory of relevant inputs and outputs of a product system;
- evaluating the potential environmental impacts associated with those inputs and outputs;
- interpreting the results of the inventory analysis and impact assessment phases in relation to the objectives of the study" (ISO 1996, p. iii).

The assessment includes the entire life cycle of the product, process, or activity, encompassing extracting and processing raw materials, manufacturing, transportation, distribution, use, re-use, maintenance, recycling and final disposal (SETAC 1991).

Further on, the LCA addresses environmental impacts of the system under study in the general areas of ecological consequences, human health and resource use (ISO 1996). It typically does not address economic considerations or social effects. Additionally, like all other models, LCA is a simplification of the physical system and cannot claim to provide an absolute and complete representation of every environmental interaction (SETAC 1993).

"The prime objectives of carrying out a LCA are:

- to provide a picture as complete as possible of the interactions of an activity with the environment
- to contribute to the understanding of the overall and interdependent nature of the environmental consequences of human activities; and
- to provide decision-makers with information which defines the environmental effects of these activities and identifies opportunities for environmental improvements" (SETAC 1993, p. 7)[4].

2.2.1.2.2 Phases of an LCA

As recommended by ISO (1996) and SETAC (1993), LCA studies should be carried out in four main steps. These four main phases, sometimes referred to as the four I's in literature, are (Figure 2.3):

- Goal and scope definition[5],
- Inventory analysis,
- Impact assessment and
- Interpretation[6].

---

[4] As to how this can be most effectively done in companies, under which conditions and limitations, and by using which applications is precisely the subject of this volume.

[5] The term "Initiation", which completes the list of four I's, was introduced by the Society for the Promotion of LCA Development (SPOLD).

[6] Formerly, SETAC called this phase "Improvement assessment", whereas ISO 14040 refers to it as "Interpretation", pointing out that the development of conclusions do form an integral part of an LCA study and should be included in this phase.

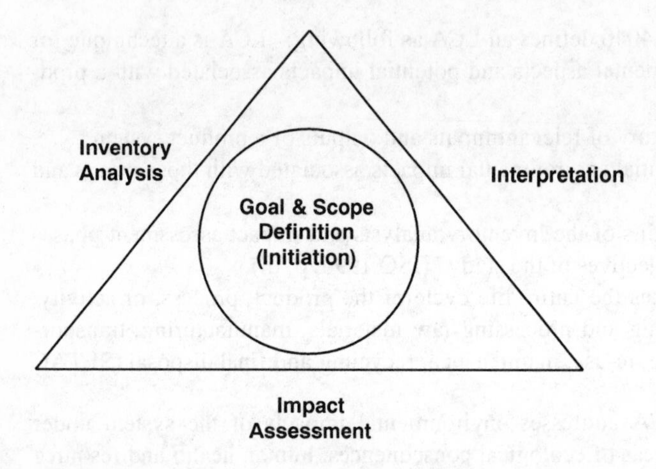

Fig. 2.3. The four phases of an LCA study ("the four I's") [adapted from SETAC 1993 and ISO 1996]

Goal and Scope Definition

In the first step of an LCA, the *goal* and the *scope* of the study have to be clearly defined and agreed upon with reference to the application(s) intended. Therefore, the goal of an LCA shall include motivations for the study, intended applications and audiences, initial data quality requirements and type of critical review (ISO 1996).

Furthermore, it is important to define the system in terms of its functional service and its boundaries. This is the subject of the scope phase, which should also include the method of impact assessment and subsequent interpretation, the data requirements, all assumptions made and the limitations (if known). All these parameters are defined according to the stated goal of the study and should be clearly stated, comprehensible and transparent. They should also indicate the representativeness of the system (and of its boundaries) in terms of technology, geography, time, market, data and data sources. In comparative studies, the equivalence of the systems being compared shall be evaluated before interpreting the results.

The appropriate definition of all these parameters at a very early stage of the study is of crucial relevance for effectively-conducting the whole LCA. Ambiguous goal and scope settings at this stage can lead to confusing results, misleading interpretations, and can cause relevant losses of time and money.

In its turn, the scope should be sufficiently well defined to ensure that the breadth, depth and detail of the study are compatible and sufficient to address the stated goal. However, it is important to realise that LCA is an iterative technique.

Therefore, if needed, the scope of the study may be changed during the study as additional information is collected (ISO 1996).

Inventory Analysis

The Inventory Analysis phase provides a comprehensive view of the flows of materials, energy, water and pollutants, in and out of the system boundaries. This phase is fundamental, since its reliability will affect the complete study[7]. The issue is not a trivial one, since many product life cycles imply both complex systems and subsystems and complex energy and material flows. However, there are precise guidelines (i.e. SETAC 1993) and ISO standard 14041 for LCA practitioners on how to take key decisions related to the definition of the systems and their boundaries, the definition of the functional unit, the data collection and calculation procedures, particularly as far as energy accounting[8] and allocation rules[9] are concerned.

Again, ISO 14040 points out that the process of conducting an inventory analysis is iterative. During this "learning by doing" process, new data requirements or limitations may be identified. This might require changes in data collection procedures and/or even revisions to the goal or scope of the study itself.

Impact Assessment

Both Goal Definition and Scoping and Inventory Analysis phases are well defined, understood, and coded (e.g. in ISO 14040, ISO 14041, SETAC 1993)[10]. On the other hand, the Impact Assessment while conceptually defined, is much less developed in terms of practical guidelines. The ISO standard 14042 is in progress and is expected to be accepted 2000.

The impact assessment aims at understanding and evaluating the environmental impacts based on the Inventory Analysis, within the framework of the goal and scope of the study. Of course, many of these impacts are context-specific and cannot be generalised. Different eco-systems might be particularly sensitive to spe-

---

[7] Unfortunately, several LCAs carried out in the past were unreliable because of inaccurate or missing data elaboration during the inventory phase (Ayres 1994).

[8] The data on energy consumption should take into account several parameters, such as types of fuel and electricity used, the efficiency of conversion of energy and the environmental inputs and output associated with the generation and use of that energy.

[9] Allocation issues derive from systems involving multiple inputs and/or multiple outputs (e.g. generation of coproducts, waste treatment processes and recycling). The allocation rules might depend on the goal and scope definition of the study, but they should be clearly stated. All assumptions made should be fully transparent and comprehensible.

[10] We highlight that the process of carrying out an LCA is well coded. Of course, the problem of availability and quality of data still remains. This is one of the main current limitations of LCA, as also shown by the results of the survey (see chapter 4) and the case-studies (see chapter 5).

cific pollutants, thus the same process might have very different environmental impacts in two different places.

Methodologically, the Draft International Standard (ISO/DIS) 14042 for impact assessment includes three *mandatory* sub-phases:

- Selection of impact categories, category indicators and models,
- Classification and
- Characterisation.

In the first sub-phase, categories should be identified and results of the inventory phase should be assigned to the impact category.

In the second sub-phase, resulting data from the Inventory Analysis are classified according to a number of impact categories. General impact categories are:

- resource depletion,
- human health and
- ecological impacts.

Specific impact categories such as global warming, acidification or eutrophication may then be identified. The selection of impact categories is goal-dependent.

In the characterisation phase, impacts in each of the selected categories are analysed, quantified and calculated. For this purpose, scientific knowledge about environmental load-response relationships is needed. Furthermore, models to represent potential and uncertain effects or impacts are needed. Inventory data alone are not a measure of environmental disturbance, as they do not provide data on exposure.

Currently, various modelling approaches are used, e.g.

- the use of equivalency factors such as ozone depletion potential, and
- the use of toxicological data, such as non-observed effect concentrations.

Beside the three mandatory sub-phases of an impact assessment, the ISO/DIS 14042 introduces *optional* elements, namely normalisation, grouping, weighting and data quality analysis. These optional sub-phases discuss and judge the level of environmental disturbance. Clearly, this is a subjective procedure. Not only environmental risks are debated; some advocate the consideration of other complex social impacts, and the need the product itself might be called into question in this phase (Fussler n.y.). Therefore, any qualitative or quantitative valuation must be as transparent as possible. All factors and assumptions used, leading to a decision on environmental preferences, should be justified and explained.

Clearly the stage of Impact Assessment is still under progress and still requires significant methodological improvement. Until now, there are no generally accepted methodologies for consistent and accurate inventory data with specific potential environmental impacts (ISO 1996).

Interpretation

In the Interpretation phase, the results of the Inventory Analysis or the Impact Assessment, or both, are organised into results that are comprehensible for decision makers. These findings may be presented as conclusions and recommendations, consistent with the goal and scope of the study. Again, drawing conclusions may involve an iterative process of reviewing the scope of the LCA as well as the nature and quality of the data collected.

It is worth noting that ISO 14040 explicitly mentions that the Interpretation of results may concern the findings of the Life Cycle Inventory only (always according to the defined goal and scope) and does not necessarily require an Impact Assessment.

Although in many cases improvement assessments based on Life-Cycle Inventory data are being performed, formal systematic procedures for this phase are not yet fully established. However, this is the most productive phase for a company, since the global view of the product life-cycle provides new insights and improvement opportunities. It is worth mentioning that all real improvement opportunities absolutely depend on the quality of the Life Cycle Inventory Analysis.

Finally, all LCA studies should include an appropriate Reporting and Critical review phase (SETAC 1993, ISO 1996). This is particularly important if the results of the LCA are to be communicated to any third party other than commissioner and practitioner of the study. ISO precisely defines the criteria for third-party reporting. It also clearly indicates the need for critical reviews, their general description as well as the guidelines for conducting critical review process itself.

2.2.1.2.3 Key features and limitations

Some major key-features of the LCA methodology are summarised here (ISO 1996):

- "LCA studies should systematically and adequately address the environmental aspects of product systems, from raw material acquisition to final disposal.
- The depth of detail and time frame of an LCA study may vary to a large extent, depending on the definition of goal and scope.
- The scope, assumptions, description of data quality, methodologies and output of LCA studies should be transparent. LCA studies should discuss and document the data sources, and be clearly and appropriately communicated.
- Provisions should be made, depending on the intended application of the LCA study, to respect confidential and proprietary matters.
- LCA methodology should be amenable to the inclusion of new scientific findings and improvements in state-of-the art technology.
- Specific requirements are applied to LCA studies which are used to make comparative assertions that are disclosed to the public.
- There is no scientific basis for reducing LCA results to a single overall score or number, since trade-offs and complexities exist for the systems analysed at different stages of their life cycles.
- There is no single method for conducting LCA studies. Organisations should have flexibility to implement LCA practically as established in this International Standard, based upon the specific application and the requirements of the user" (ISO 14040, p. 3f).

On the other hand, there are important *limitations* of LCA that need to be mentioned in advance.

- The most important one is that any LCA will necessarily involve assumptions and subjective valuation procedures. These assumptions must be fully communicated. As the results might differ substantially according to the assumptions made, great caution should be used in making environmental claims (i.e. in the comparison of products for the public) based on LCA.[11]
- A second big problem is the availability and quality of data. Several LCAs carried out in the past proved to be unreliable because of inaccurate or missing data elaboration during the inventory phase (Ayres 1994). In particular, there was no use of systematic verification methods (i.e. mass-balance verification), which have only recently been introduced in some data-base. Very clearly, the issue of the quality of data, of how to find missing data and/or correct unreliable data still requires a lot of methodological and scientific work. As reported in the case-studies, several strategies are now being looked at to tackle this problem. One possible path seems a collaborative attitude within sectors and countries and the creation of public data-banks.

  As far as this is concerned, it is important to report that at a recent workshop of the Comité Européen de Normalisation (CEN) on pre-normative research on LCA, it was decided to establish new working groups focusing on LCA data format exchange and on sector-studies. The latter will tackle with the issues of data availability and quality, confidentiality, and simplification rules, recognising that these can be significantly different from each other according to the specific industry sector (CEN 1999).
- Finally, it should be clearly understood that LCA is only one of several environmental management tools and might not always be the most appropriate one in all situations. Decisions for action in a company typically involve other factors such as risks, benefits, costs, which include technical, economic, and social aspects, which are *not* addressed by LCA.

2.2.1.2.4 Screening or streamlined LCA

A full LCA according to ISO 14040 is quite a time consuming and expensive task. In the last years discussion about simplifying and/or streamlining LCA has arisen for several reasons, e.g:

- some decisions have to be made before the results of a full LCA are available (business cannot wait);
- many companies, above all SMEs, want to make decisions on the basis of a life cycle approach, but are not able to conduct a full LCA internally (in our case-studies we found out that this is a major factor of failure);

[11] Indeed, one of the most important research results of our case-studies is that current methodological state-of-the art LCAs cannot be used for marketing purposes.

- a company producing different products by similar processes will not work on an LCA for every product, but wants to be able to make decisions on the basis of modifying slightly a few core studies;
- often, data are not or hardly available for different specific stages of a life cycle of a product, so that a full LCA is not always possible.

For these reasons the term "screening or streamlined LCA" has been recently adopted by SETAC (1997).

By screening LCA SETAC defines a procedure that identifies some particular characteristic or key issues associated with LCA, which will normally be the subject of further, more intensive study. Such key issues can be:

- the most important environmental inputs/outputs or environmental effects associated with the particular life cycle,
- the stages of the life cycle giving rise to the most significant environmental inputs or outputs,
- the occurrence of major gaps in the available data.

Streamlining LCA is understood as a simplification of an LCA which can be done by the exclusion of certain

- life cycle stages,
- system inputs or outputs or
- impact categories.

A further simplification is possible by using generic data modules rather than specific data for the system[12].

The simplification procedure consists of two elements:

- Screening: identification of elements of an LCA that can be omitted or where generic data can be used without significantly affecting accuracy or the final result
- Exclusion: removal of the identified elements to procedure the simplified LCA

For the screening step several quantitative and semi-quantitative methods are proposed, where material intensity, energy consumption or key substances are taken as parameters. In all cases a sensitivity analysis should be carried out to test proving the correctness of the simplification procedure.

In the near future, further efforts will be made to develop a generally accepted method for reducing the time consumption and costs of LCAs. This reflects the fact that on one hand there is a large interest to implement the ideas and approach of LCAs, but on the other hand resources are often limited, particularly in SMEs. We feel that one of the most important steps in this direction would be to compile generic data modules of the sort noted above.

[12] Again, on the basis of both the survey and the case-studies, we stress the importance of availability and quality of data. Possible strategies to tackle this issue include creating public data-banks, other support mechanisms for SMEs, collaborative branch studies.

2.2.2 LCA applications[13]

LCA – by allowing the identification of product weaknesses and positive attributes in comparison with company standards, competitors products, labelling criteria or alternative services – assists company managers in their decisions. The criteria generally used in the literature (Smith et al. 1998; The Nordic Council 1992; FTU/VITO/IÖW 1995a and 1995b) to classify LCA applications distinguishes between internal and external[14] applications of LCA:

- for internal purposes: LCA helps management in improving product environmental quality and making strategic decisions;
- for external purposes: LCA helps in improving marketing choices, joining labelling schemes and informing various stakeholders.

Another classification (Weidema et al. 1993) considers three different questions relevant to the environmental management of products:

- what is the status of the current programme?
- what are the corporate vision and policies?
- what are the actions necessary to implement policies?

These questions can be addressed also by using LCA during specific management phases, namely:

- use of LCA to assess the status of current programme, i.e. for:
  - the environmental evaluation to ensure product compliance,
  - the evaluation of existing product development process with life cycle considerations.
- use of LCA to support corporate vision and policies, i.e. by defining:
  - strategies to improve energy and material management,
  - strategies to minimise solid wastes,
  - strategies to preclude environmental damage.
- use of LCA to support actions to implement policies, i.e. by:
  - developing standards world wide,
  - fostering education and training programmes.

The classification proposed here takes into consideration the different approaches mentioned above (especially the internal vs. the external use of LCA), and includes an *allocation* of different applications of LCA to specific steps in the product development chain.

Figure 2.4 shows a simplified product development cycle that includes four steps: strategic decision, R&D and design, procurement and production, and marketing. The four steps have to be considered as a functional sequence but it has to be considered that the whole process of product development has several feed

---

[13] An exhaustive description of the applications of LCA by business in Switzerland, Germany, Italy and Sweden is presented in chapter 4; see also Baumann (1996) and Grotz/Scholl (1996).

[14] The terms internal and external are related to company perspective and to the different users of the information provided by the LCA.

backs (e.g. on one hand company strategies determine all the other steps, on the other hand market demand and expectations can significantly influence the strategies, etc.).

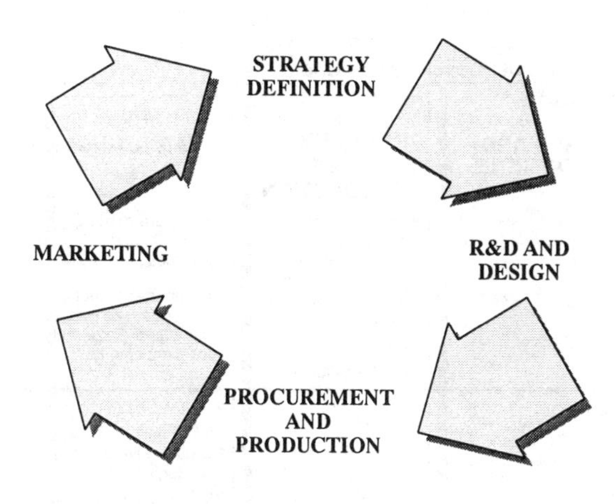

Fig. 2.4. Product development chain

Figure 2.5 shows in detail a possible classification of LCA applications according to the product development chain and to the direction of use (internal vs. external). The following parts of the subsection 2.2.2 are based on this classification. A first distinction can be made with respect to the time-scale of the use of LCA results. LCAs used in the strategy definition and/or in the R&D and design steps[15] are more prospective studies, that is they are concerned with the development of new products, and the comparison of current products with possible alternatives.

[15] The distinction between the application of LCA for R&D and design and for the definition of some specific company strategies concerning product environmental innovation can indeed be very subtle. In our meaning, if the product innovation changes suggested by an LCA are rather marginal (e.g. optimisation of production process, choice of alternative materials) we refer to the use of LCA for R,D&D. On the contrary, if the suggested changes are more significant, we refer to it as LCA for product-related strategic decisions (i.e. radical changes in the product life-cycle, shift from product to service, etc.).

In these cases, the results of LCA are meant to be used rather in the mid-long term. On the contrary, in principle the application in the marketing step and to some extent in the procurement step might be immediate and can be slightly more retrospective (for learning and confirmation purposes).

The application of LCA along the four main steps of the product development chain is described in subsections 2.2.2.1 to 2.2.2.4; the application of LCA for information, training and education is discussed in subsection 2.2.2.5.

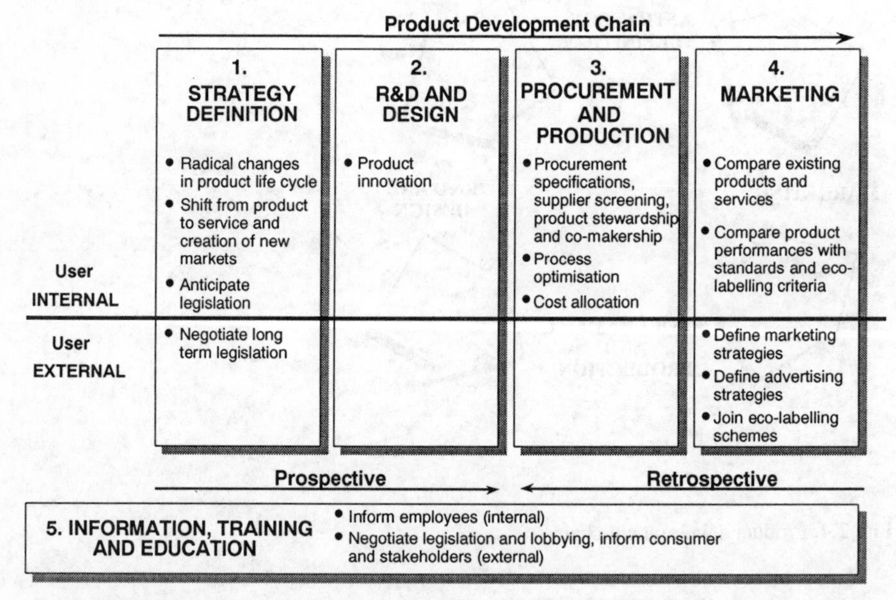

Fig. 2.5. Applications of LCA in the product development chain

2.2.2.1 Strategy definition

The goal of a company is certainly not only short-term profit[16]. It is also primarily to increase the real service provided to society as a whole and to increase the satisfaction of customers, stockholders and the public in general in the long run. All these are crucial factors for guaranteeing the survival of the firm itself over time.

To achieve these objectives, explicit recognition at the policy-making level of the firm is needed, that the earth's resources are finite, its carrying capacities are limited, and that firms bear some responsibility for not making problems worse in this regard. This is the basic concept which lies behind eco-efficiency, as sup-

[16] The teaching message in some famous business schools is that the goal is to increase "shareholder value".

ported by the World Business Council for Sustainable Development (WBSCD 1995): „Eco-efficiency is reached by the delivery of competitively priced goods and services that satisfy human needs and bring quality of life, while progresivelòy reducing ecological impacts and resource intensity *throughout the life-cycle*, to a level at least in line with earth's estimated carrying capacity".

In operative terms, this implies firms to define long-term strategies taking into account both existing and likely future environmental burdens and risks (and costs). Incidentally, reducing likely costs of business in the future environment will be a real saving for the firm.

In the long run it is very important that these environmental improvement criteria intervene at the very early stage, especially with respect to decisions about product design changes and production. To obtain this, a systems (life-cycle) viewpoint with respect to both processes and products is needed.

Example

Recently, the Swiss Telecom PTT introduced a new chip-card for public phone calls, in order to replace the call system with coins. An LCA was made to assess the environmental implications. At the beginning, the chip-card was introduced as a single-use product. Therefore, the recycling of the used cards was a major focus of the study. However, the LCA showed firstly that the production of a chip-card is quite energy consuming and includes some polluting production steps. Moreover, the study also showed that additional environmental burdens would be added by recycling. On the contrary, the LCA suggested that a real over-all environmental improvement would come by the repeated use of the card, by recharging it. So, Telecom decided not to introduce an expensive recycling system, but rather to enforce the development of rechargeable cards and to introduce it as soon as possible. This example shows that an LCA can be used as an instrument for changing management strategies (not necessarily on the long-term). The example shows also two other significant aspects of LCA: the first one is the importance to focus on the actual *final service* provided and not only on the product itself; the second is the importance of taking into account the whole life-cycle and not focusing on the final waste step only.

2.2.2.1.1 Radical changes in product life-cycle

The main motivation for the consideration of radical changes in a product's life cycle is the explicit recognition that obtaining further environmental gains[17] via end-of-pipe approaches is becoming extremely, sometimes unaffordably, costly. On the other hand, it is clear that there are often many opportunities for reducing the environmental burden at lower costs (or even at profit) through process change

[17] Independently of the motivations why these gains are to be achieved, whether external (to respond to mandatory regulations or to face to environmental pressure from the public) or internal ones.

or product innovation. Some of these opportunities have already proven to be cost effective. Many others remain to be exploited.

LCA can be very effective as a supporting tool to formulate mid-long term strategies implying radical product life-cycle changes, by identifying:

- bottlenecks of current production and product life-cycle,
- possible alternative production processes and
- better design opportunities, which can facilitate the re-use, the maintenance and repair and the re-manufacturing of the product, and the recycling of materials.

2.2.2.1.2 Shift from product to service

Companies might change the nature of their business, going far beyond product innovation, both as far as new products, product design and/or production processes are concerned. As discussed in the last years, (e.g. in the WBCSD Antwerp Eco-Efficiency Workshop [WBCSD 1995], OECD 1998a), the eco-efficiency concept might imply the shift (over time) away from producing material goods to selling (immaterial) services. This fact can have tremendous (positive) implications on the environmental impact generated by a firm. In many cases, this approach can lead to "win-win" situations, reducing environmental burden and increasing profit for the firm (see examples below).

Examples

Several examples of this paradigm shift exist. We can mention Integrated Materials Management at Cookson PLC, leasing of copy-machines at Xerox, renting tires by some specialised companies. Other examples come from Dow which has developed a solvent-rental business for electronic industry and from Ciba-Geigy that has shifted its former basic approach in the agro-chemical sector from the simple sale of chemical products to the supply of integrated pest management services.

In principle, the support of a shift from product to services would be one of the most interesting prospective applications of LCA. However, we have to mention that our case-studies showed that at present none of our interviewed companies is using LCA for this specific goal today nor plans to do so in the near future (see chpater 5).

2.2.2.1.3 Anticipate and negotiate legislation

The public and political pressure (from different stakeholders) to modify business accordingly to mandatory regulation will inevitably increase. These constraints (both physical and institutional) will become more and more part of business life. Life Cycle Thinking and Life Cycle Approaches can be very useful for planning scenarios which anticipate the change of external constraints. There are several motivations for firms to anticipate legislation instead to just wait for it. Some major ones are:

- the possibility to plan better and to gain flexibility in scheduling and the choice of technology;
- the risk of losing "green" customers and of harming the company's reputation by a bad environmental performance.

To some extent LCA can also influence the change itself, as companies can support public administrators for creating a level playing field by means of appropriate regulation and tax policy.

2.2.2.2 Research, development and design

This is the stage where the LCA approach is considered to be the most promising one by companies. As showed by our case-studies (see chapter 5), in companies where LCA has reached a mature phase, R&D and product design are normally assisted by LCAs, in particular in the choice of materials and energy sources.

Examples

Car producers have been using LCA as a R&D instrument for a long time. Volvo, FIAT, Volkswagen, BMW and Mercedes are just some examples in this field. At least an LCA inventory is used at the R&D stage by large companies in the electronic appliances sector (Electrolux, Merloni, Philips). Some LCA applications are registered also in the food industry (for example food division of Unilever). Our case-studies (see chapter 5) provide several examples.

2.2.2.3 Procurement and production

2.2.2.3.1 Procurement specifications, supplier screening, product co-makership and stewardship

Traditionally, supplier screening policies have used two main criteria: cost and product performance. However, clients are increasingly asking suppliers to comply with quality and environmental criteria. ISO 9000 is becoming a major standard in public and private procurement. Similar standards, addressing environmental issues (like EMAS regulation and the ISO 14000 series), are supposed to play an important role in the near future[18]. These kinds of procurement policies, based on qualitative information, are mostly due to the need for creating, preserving or improving a green image.

Other approaches can involve the use of quantitative information (such as data on impact and physical flows) from and to the environment.

---

[18] Volvo has recently organised a conference inviting 700 suppliers to illustrate EMAS regulations and to favour the adoption of the scheme by all of them.

LCA, by providing information on the most critical bottlenecks and improvement opportunities in the product life cycle can suggest priority actions and assist management in:

- Optimising production processes to minimise resource- and energy-use and emissions.
- Defining procurement specifications to be followed by the purchasing department and therefore by suppliers. This activity has the aim of continuously controlling the quality inputs and pushing current suppliers to improve their product and processes on the basis of defined standards.
- Screening suppliers on the basis of the environmental performance of their processes or products. This activity can be a stand alone exercise and has the aim of selecting suppliers on the basis of their environmental performance (of course other criteria will be examined as well).
- Optimising logistics, undertaking take-back policies, etc.

The effectiveness of these instruments depend on many factors:

- power and position of the company along the product value chain,
- shared responsibility among different actors along the product value chain,
- level of competition among suppliers.

A step further in procurement policies is represented by the so called product stewardship[19] and product co-makership. These approaches are conceptually based on shared responsibility and co-operation among suppliers, producers and clients along the value chain of a given product. This approach, poorly implemented up to now, is considered favourably by policy makers and companies because it could lead to a cost-effective reduction of product overall environmental impact.

Examples

An example is given in the textile sector by the German company Kunert. As an important customer of the German AKS Kammgarnspinnerei, Kunert asked AKS to implement changes in their production processes. For AKS it was their first ecological engagement. The interesting aspect to be highlighted here is that the LCA led to results, which were quite interesting for AKS and which influenced several choices of the management in *other* areas of improvement. This example shows a possible kind of co-operation and shows how the life cycle approach can provide advantages for several subjects along the procurement chain.

A second example is given by green public procurement. In Switzerland, several governmental authorities use this method for improving buildings in an ecological way. In particular, they require an ecological assessment of paint substances by means of a method called BUWAL 1992. The method is based on a

[19] This concept has been used in the chemical industry since 1987.

life-cycle approach and was commonly elaborated by three Swiss governmental authorities, an association of producers and an association of final users[20].

This example shows how an association of producers can use LCA together with governmental offices as an instrument for an objective and neutral evaluation of its products. This is also an example of shared responsibility between manufacturers and clients.

2.2.2.3.2 Cost allocation

Environmental flows and impacts are costly both for the company and the society. A company bears only part of the costs related to processing and production. A large amount of environmental costs generated by a given product or process are born by the public[21].

Environmental costs can be divided into three main categories:

- costs conventionally calculated include end-of-pipe measures and those costs conventionally considered by company management accounting systems as environmental costs;
- internal costs is a broad category that includes the conventionally calculated costs and all those costs that, even if borne by the company, are not considered to be related to environmental protection; this category includes hidden costs, non tangible costs and potential future costs;
- social costs are represented by all the costs, direct and indirect, caused by a production activity that are born by the community.

An inefficient product life cycle management can cause both external and internal costs:

- external costs can be caused by excessive emissions during raw material acquisition, transportation, distribution, use and disposal, land use changes, soil erosion, etc.
- internal costs include those hidden costs related to product life cycle.

Today a company should be interested in identifying and reducing internal costs related to product life cycle and, as a proactive approach, in identifying at least the direct external costs[22].

---

[20] The involved subjects were respectively: the Federal Office for Environment, Forest and Landscape [BUWAL], the Office for Federal Buildings [AfB] and the Federal Office for Public Health [BAG]); the Association of Swiss Varnish and Paint Manufacturers; and the Swiss Master Painters and Plasterers Association.

[21] Economists are used to call them externalities.

[22] In a few cases it is possible that an improvement at a certain stage can cause a worsening at another stage of the life-cycle. In an even worse case the increase of environmental burden could be higher than the improvement due to the fulfilment of eco-label criteria. It is important to notice that several proposed methods to calculate external costs do exist (i.e. cost of restoration, willingness-to-pay), however there is not a single, generally-accepted method. The issue is particularly difficult when it refers to an environmental risk rather than an actual damage (e.g. the problem of global warming).

The second activity, as shown by some relevant business experience, can be seen as the anticipation of new legislation or simply of stakeholders pressure towards increased producer responsibility along the life cycle.

In principle, LCC is able to combine information (like impacts of physical flows) related to LCA with cost figures. LCC can be combined with LCA or represent a stand alone instrument based on life cycle approach where environmental costs related to purchasing, transportation and distribution are identified and allocated to specific life cycle steps.

Examples

Some examples of LCC are reported in the field of computer and hi-tech appliances industry: AT&T has been investigated by EPA for its ongoing LCC policy, Rank Xerox is focusing on the toner cartridge take back policies and their impact on company cost structure.

Car manufacturers are using a combination of Life Cycle Thinking and cost identification for their take back programmes for cars at the end of their life cycle. An example of LCC focused on external cost comes from Ontario Hydro (Canada) that is assessing externalities related to hydro energy life cycle.

2.2.2.4 Market

2.2.2.4.1 Comparison of existing products and services

A possible application of LCA, done by a producer, is the comparison of his own products with each other and with those of his competitors. This application of LCA gives a kind of benchmarking of the own products with respect to the market situation and can eventually provide some ideas for improvement of production processes and product design. A lot of LCA-studies have been done within this field, but only a few have been published. Mostly, the main goal of LCA in this case is for learning. The results might be used internally for redesigning the products and processes but they are usually not communicated to the public. Rightly so, because while there are no constraints in using comparative LCA for internal goals, there are clear limitations and rules (defined by ISO) to be followed if the results of comparative LCA studies are to be communicated to the public (see subsection 2.2.2.4.3).

Apart from the use of the results, monitoring competitors can be a strong driver for starting LCA. As a matter of fact, in our case-studies we found that it is an important factor in specific stages of LCA adoption.

Example

One of our case-studies (Henkel, see subsection 5.2.2.3) was motivated to respond to LCA activities of an important competitor (Procter & Gamble).

2.2.2.4.2 *Assessing the gaps from the requirements of an eco-label*

Often, the conditions for elaborating requirements for an official eco-label for a certain product group are set up on the basis of single criteria. The latter are focusing on the main environmental critical points, which are usually referring to a single step of the product life-cycle. Examples for such criteria are CFC free sprays, paints with low solvent content, paper made from recycled paper to a certain extent, brake-linings without asbestos, packaging for transportation with several uses and so on. In this case, companies carrying out an LCA can use part of the results to assess their gap from eco-label requirements (an important spin-off result).

However, it should be highlighted that the results of an LCA, giving a broader perspective, might be also used by companies in a critical way with respect to eco-label criteria (and to legislation)[23].

2.2.2.4.3 *Definition of marketing and advertising policies*

Marketing and advertising policies are an important part of company attitude towards the environment and reflect strategies, programmes, objectives and achievements of the company. For the application of LCA results in advertising, it is however very important that they are done according to strict guidelines[24] because a comparative use of LCA methodology can lead to wrong interpretation of results and advertising would not be very reliable in that case. In some cases this can easily lead to the misuse of an LCA where usually only the advantageous part of the LCA is published.

Many companies have had high expectations concerning the use of LCA for marketing purposes, especially in Germany. However, one main result of our research shows that at the current methodological state-of-the-art level, companies do not use LCA directly for marketing (see section 5.3).

This general result needs some further discussion though. A first distinction has to be made between business-to-business and business-to-consumer relations. While in the latter case all the above mentioned limitations hold strictly, in the first case, companies have the possibility to carefully describe the results of LCA and its underlying assumptions to their clients. A second way which allows the

---

[23] In a few cases it is possible that an improvement at a certain stage can cause a worsening at another stage of the life-cycle. In an even worse case the increase of environmental burden could be higher than the improvement due to the fulfilment of eco-label criteria. An example are waterbased paints, which fulfil an important criteria for different ecolabels, namely they decrease the burden of volatile organic compounds (VOC). On the other hand, it is possible that, for a building renovation, these paints have to be removed with solvent based agents or by using a huge amount of thermal energy. In that case the advantage of this paint will be cancelled. For this reason it is helpful to look at the whole life cycle and to base the eco-label criteria on a LCA.

[24] As a matter of fact, ISO 14040 codes both the phases of third part reporting and critical review. The standard 14040 has also specific requirements for the application of an LCA-study for a comparative marketing, namely the critical review (type III).

application of LCA for marketing and advertising is for providing quantitative information in environmental product declarations (ISO-type III). There are some example of this kind of application in Nordic countries.

Examples

Traces of Life Cycle Approach can be found in the advertising of the Audi A8 model, a car characterised by its aluminium body coachwork. This advertising points out the environmental benefits of the car explaining the bauxite extraction, aluminium production, car use and disposal. In particular the message explains that the choice of the material reduces the weight of the car (by a given percentage), improves energy efficiency (figures are given), increases the car life, and allows a 100% recycling of the raw material. The picture, rather than showing the car in itself relates to the bauxite mineral. The message also reports on the higher price of the car, which is supposed to be offset by better energy performance and by the pleasure of owning a highly environmental performing product[25]. This is an example of correct application of LCA for advertising.

On the other hand, they do exist also examples of misuse of LCA. One is given by a comparative study carried out in Switzerland in the past comparing PVC with wood and aluminium window frames. The study was done according to the guidelines, but in the communication to the public only some results were presented, showing less environmental impacts from the PVC-window. This communication was based on a high recycling rate for PVC, which was only a goal for the future and has not yet been reached. This is a clear example of the misuse of LCA in advertising and marketing. Today, this would not comply with the ISO 14040 standard.

2.2.2.5 Information

2.2.2.5.1 Internal information

LCAs can be used as an instrument for internal information. An important motivation for this is monitoring and imitation of competitors (a company which does not know the methodology might fear a disadvantage on the market or a lack of competence in discussions within the branch). However, there is also a second more internal goal related to the satisfaction and motivations of employees.

In all our case-studies the value of learning from LCA activities is stressed (see section 5.3). In general, the level of satisfaction expressed within Engineering, Design and general technical departments with respect to LCA is very high.

---

[25] We point out however, that one of our case-studies – FIAT – comes to almost opposite conclusions (at the present boundary conditions of secondary aluminium market).

2.2.2.5.2 Information to consumers and stakeholders

The increased attention of stakeholders and consumers toward the environment is potentially one of the main factors for companies to adopt LCA and Life Cycle Approaches. As mentioned before, the strict limitations holding for LCA application in comparative marketing, do not exist if LCA is used for generic information and education of consumers and stakeholders. Firms can make a major contribution in educating society, both directly and indirectly. To some extent, this can also happen through the channel of public administration. However, this can best be accomplished through NGOs, trade unions & consumer associations, for the very simple reason that these groups can often communicate with final customers much more effectively than companies themselves. In their turn, well-educated consumers create a green market, thus further pushing the whole business community towards more environmentally-sound products and/or services.

A Life Cycle Thinking / Approach is a key factor for establishing the required level of confidence between the company and the above mentioned external groups, thus fostering multi-stakeholder initiatives for implementation.

Examples

The results of the survey indicate that while LCA are presently not used for marketing, they are heavily used for internal information and some kind of external information and education to consumers and stakeholders (see chapter 4).

An example is Greenpeace and Brent Spar: Greenpeace UK had to apologise to Shell, as the latter showed with an LCA that the disassembling of the Brent Spar was actually much worse for the environment than its disposal in the sea.

3 Application of LCA in general

The history of product-oriented environmental tools goes back more than twenty-five years. Amongst the first studies – not yet called LCA – was a study carried out in the United States in 1969 on behalf of Coca-Cola and another in 1974 in Germany on behalf of the German Ministry of Research (Oberbacher et al. 1974)[1].

The circle of potential users and appliers of LCA can be as widespread as was mentioned in the previous chapter. Several reports refer to the broad field where LCA is already in use. However, in this chapter, we will refer to the four European countries which joined the study as a whole. We present here a survey on the application of Life Cycle Assessment by all the different users and actors in these countries; the state of this information is as of the end of 1996.

The following sections analysis look for year of publication, commissioning body, business sector(s) involved, size of commissioning enterprise and products investigated[2].

3.1 Methods

The number of applications of LCA in the four countries considered (Germany, Italy, Sweden and Switzerland) refers both to ongoing and completed LCA studies. Information and data have been collected through:

- mailing questionnaires to selected enterprises, business associations and institutions,
- contacting with researchers, consultants and other key people,
- analysing of existing bibliographies (Grotz/Rubik 1997, Ambiente Italia's database),
- examining of relevant journals and publications,

---

[1] Boustead (1996), Fink (1997) and Oberbacher et al. (1996) report on the history on LCA in several countries.

[2] Most LCA researchers do not restrict the term "product" to material products, but also include processes and services.

- using some databases elaborated in former projects of members of the project team[3].

In addition, we refer also to other surveys which have been carried out and published.

3.2 General application

Table 3.1 presents the total results of known LCA studies in the four countries. In Germany, almost 300 studies have been compiled or are in their planning stages. In Switzerland and Sweden some 150 studies have been recorded, whereas Italy, by contrast, lags behind[4].

Table 3.1. Known LCA studies in various countries (as of end of 1996 – with the exception of Italy)

Studies	Switzerland	Germany	Italy	Sweden
• Completed studies	149	250	55	137
• Current studies	n.a.	36	n.a.	6
• Projected studies	n.a.	2	n.a.	2
Total	**149**	**288**	**55**	**145**
LCA-studies per 10 billion GDP (in US-$ for 1996)	6.42	1.55	0.57	7.25

n.a. not available

However, we could also take into account the different economic importance of these studies within the four countries. Considering the GDP of the four countries, Sweden is the most important LCA-country, followed by Switzerland. In relative terms, the importance of German companies as LCA-applicants is not so great; not unexpectedly Italian companies are at the very beginning.

[3] IÖW carried out a survey together with the German BDI on the application of LCA some years ago (Grotz/Scholl 1996). Some of the results could be used for our current purposes.

In Sweden, the results of the Swedish Business Environmental Barometer were used. The Swedish Business Environmental Barometer is a questionnaire survey on environmental practice and opinions in industry carried out both in 1993 and 1995, and will be repeated every two years. It has been analysed here with regard to the occurrence of LCA. The survey has been directed at two groups of respondents: all 212 companies listed on the Swedish Stock Exchange (since 1993) and a random sample of companies (633) with more than ten employees (since 1995).

[4] However, in Italy since the end of 1996, several new LCA studies have been carried out both by the companies already performing LCA and by other companies.

It should be emphasised that the statistical data is incomplete, as a number of studies could not be included due to the commissioning bodies' request for confidentiality[5]. *The figures should be regarded as a lower limit.* An exact statistical analysis poses several serious problems:

- With regard to ongoing LCA studies, the degree of information is sometimes quite meagre. The commissioning body is known, but sometimes not a publication date.
- The commissioning bodies of some LCA studies are not mentioned in the studies themselves. These studies have been allocated to the category "Unknown commissioning body".
- Some businesses which we contacted returned completed questionnaires, indicating that they had commissioned several LCA studies. If an actual figure was mentioned, this number was counted; if an actual figure was missing (by the answer "More"), it was counted as "one" LCA study.
- Some enterprises interviewed indicated that they had completed LCA studies and also that LCA studies were under way. In such cases, we counted the study twice, once for the completed LCA study, once for the ongoing LCA study. The ongoing study was not included in the statistical analysis with respect to year of publication, commissioning body, business and subject.

The statistical information presented in this chapter should be viewed in the light of the above reservations.

3.3 Years

As already mentioned, the history of LCA goes back to the late sixties and seventies. As can be seen from Fig. 3.1, almost no LCA studies were carried out before the beginning of the1980s, but their numbers increased from the mid-1980s and gained momentum in the 1990s[6].

In *Germany*, there have been at least 16 studies a year since 1990. The rather low figure of seven studies for 1996 should by no means be interpreted as a downward trend, since at the time of this statistical survey (summer 1996) a considerable number of studies had not been completed. Moreover, there is always a time lag between completion of a study and its public disclosure.

In *Switzerland* the number of LCA studies has stood at 15-20 a year during the present decade.

---

[5] For example, by the end of 1991 the American company Franklin Associates compiled about 70 studies, of which, however, only four are known. The OECD (1995, p. 39) stated that only 10% of the Franklin studies are available to the public & only 5% of the French consultancy Ecobilan. These figures were slightly modified; accordingly 30-60% of all LCA studies commissioned by companies are for internal use only (Smith et al. 1998, p. 29).

[6] Results for Italy are unfortunately not available.

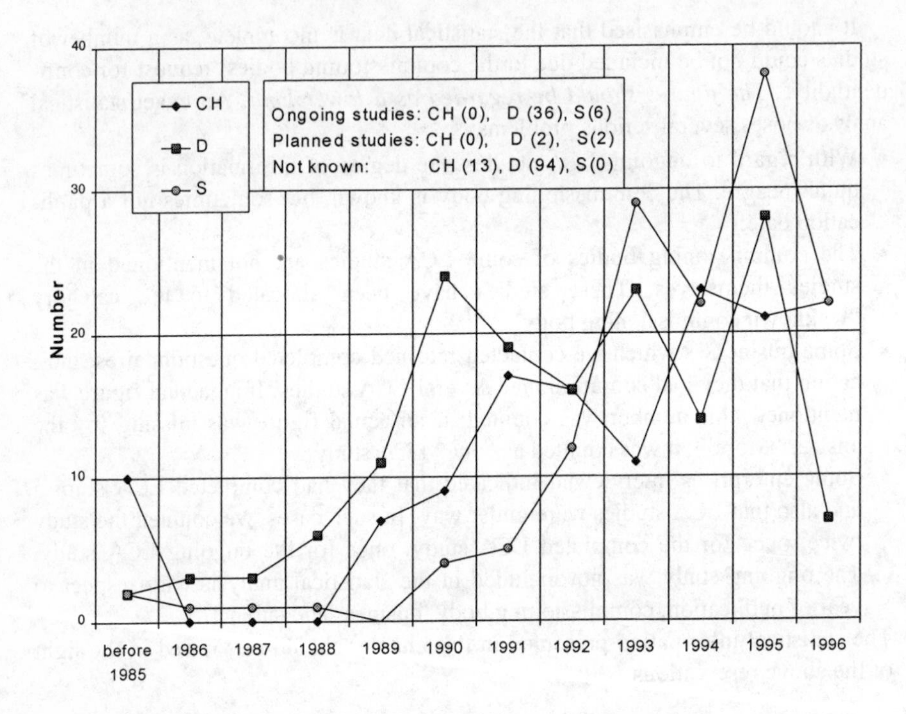

Fig. 3.1. Year of publication of LCA studies in various countries (absolute numbers)

- In *Sweden* the boom began in about 1992, since then almost the same number of studies have been recorded as in Germany. These results confirm that of the Environmental Business Barometer which indicated a slight increase in LCA use in industry between 1993 and 1995 (Baumann 1996, p. 124). The first Swedish LCAs were carried out before 1985 by a consultancy firm, Miljöbalans, established in the early 1970s.
- This type of data is not available for *Italy*.

The low count of LCAs in 1996 is probably due to projects published late in 1996 not being collected in this survey.

3.4 Commissioning bodies

LCA-studies can be carried out on behalf of different commissioners. Fig. 3.2 gives a breakdown of commissioners of LCA-studies.

- In *Switzerland*, LCA studies commissioned by companies are significant (42%); compared with Germany, the public sector is also a significant commissioning organisation.

- In *Germany*, companies commission most LCA-studies (61%) followed by studies jointly commissioned by industrial associations. The public sector is also an important commissioning organisation, although it ranks much lower than the private sector.
- In *Sweden*, the private sector commissions approximately 60% of the studies, and the public sector has a relatively insignificant share.
- In *Italy*, the private sector predominates even more strongly (85%)

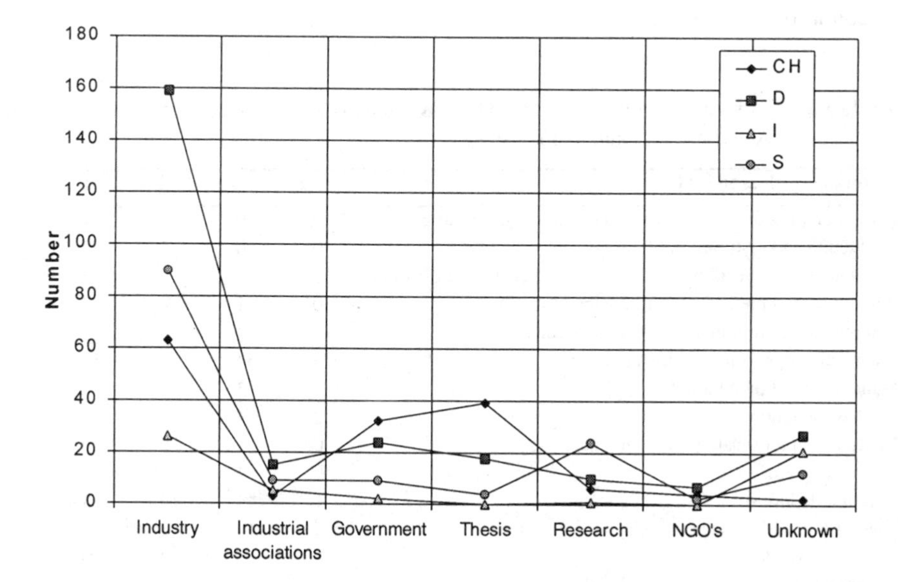

Fig. 3.2. LCA commissioning organisations in various countries (absolute numbers)

However, the role of the public sector should not be underestimated[7]. LCA studies commissioned by the public sector have repercussions on the private sector. Whereas LCA studies for the public sector attach greater importance to generic data and tend to publish "average" numbers, the private sector is more interested in and reacts to the specific positioning as opposed to the average. LCA provides a most suitable tool for this.

[7] See also chapter 6.

3.5 Business as clients

As shown in the previous section, business is the most important commissioner of LCA. But in which sectors do the LCA-activities take place? Table 3.2 classifies companies according to their primary economic activity and the European classification of industrial sectors according to Eurostat (1996):

- In *Switzerland*, the rubber and plastics processing industry commissions most LCAs (16 studies), followed by the handling and processing of non-metallic minerals/glass manufacture and processing (ten studies). All other branches account for less than 10%.

Table 3.2. LCA studies commissioned by business and classified by sector according to NACE (absolute numbers)

Sectors according to NACE	CH	D	I	S
Extraction of petroleum/ natural gas, mineral oil refining	0	0	1	1
Production and distribution of energy	0	0	1	2
Water supply; collection, purification and distribution of water	0	0	0	1
Production and preliminary processing of metals	0	1	0	2
Manufacture of non-metallic mineral products	10	4	0	1
Chemical and man-made fibres industry	5	54	8	15
Manufacture of metal articles	1	3	0	0
Mechanical engineering	0	4	1	1
Office machinery/ data-processing machinery	0	4	0	0
Electrical engineering	0	4	12[a]	20
Automobile industry	1	44	14	13
Food, drink, and tobacco industry	5	17	2	3
Textile industry	0	7	0	0
Timber and wooden furniture industries	4	1	0	6
Paper/ paper products; printing/ publishing	1	8	5	6
Processing of rubber and plastics	16	5	0	2
Other manufacturing industries	4	0	1	9
Building and civil engineering	0	1	0	2
Trade	4	1	1	2
Traffic related activities	0	0	0	1
Other regional traffic	0	0	0	1
Shipping	0	0	0	1
Aviation	1	0	0	0
Communication	1	0	1	0
Credits and loans	2	0	0	0
Not known	8	1	0	24
Total	**63**	**159**	**47**	**113**

a This figure comprises electrodomestic manufacturing industries

- In *Germany*, the chemical industry and the motor industry are the most important commissioners of LCAs. The food processing and luxury foodstuff industries are also of considerable importance. To date, all the other sectors account for less than ten LCA studies each. An analysis of companies by size (determined by number of employees) show that 84% of businesses that compiled or commissioned LCA are large-scale enterprises. Only 7% are small and medium-sized enterprises, i.e. those with less than 250 employees.

- For *Italy*, the automotive, chemical and electrical engineering industries appear to have the largest share. It has to be pointed out that more than 50% of the firms belong to multinational companies[8].
- In *Sweden* electrical engineering predominates (20 studies) followed by the chemical industry (15 studies) and the automotive industry (13 studies). The Environmental Business Barometer (Baumann 1996) showed that the forestry/pulp & paper industry often uses LCA. As in Germany, large-scale companies are to the fore.

The degree of competence and of in-house experts is often relatively important. For instance, the multinational Procter & Gamble employs a staff of 15 worldwide, whose work solely concerns LCA. Dow Chemical and Volvo employ six people in this area (Smith et al. 1998, p. 26). It is, however, not possible to make a generally valid statement on the issue of intensity.

3.6 Subjects

Compared with commissioning bodies, the subject matter or product groups covered by LCA studies are quite broadly distributed (see Table 3.3).

- In *Switzerland*, construction materials are of paramount importance (45 LCAs) followed by energy and packaging (31 and 30 LCAs respectively).
- In *Germany*, three areas, namely motor vehicles/motor vehicle components / transport, chemical products and packaging are similar in importance (between 36 and 46 LCA studies). Other significant areas are those of hygiene/cleaning and construction materials.
- In *Italy* motor vehicles (15 studies) predominate followed by packaging (ten studies).
- In *Sweden*, a lot of the known LCAs concern packaging and household wastes either as the choice between packaging materials or as the choice between waste management systems (see categories 'packaging' and ' waste'). The first ones of this kind are of an 'older' date, i.e. around 1990 and 1991 when the government's Packaging Commission was operating. Since then, packaging and management of household wastes have been issues in an ongoing debate, and the number of studies on 'packaging', 'waste' and 'paper/print' have built

[8] The importance of multinational companies in Italy is confirmed by the survey which we carried out (see chapter 4).

up, however most LCA studies were compiled in the packaging area (27 studies). Of the more recent LCAs, many concern transportation and car parts, buildings and building materials, and white goods (e.g. refrigerators, vacuum cleaners etc.). Some of the most recent ones concern complex functions and are found in the category 'others'. Examples here are LCAs of 'the office' and of 'sanitary systems'. The LCAs of the chemical industry are found in several categories, such as 'synthetic materials', 'hygienic products', 'furniture', 'energy' and 'building materials', followed by motor vehicles, construction materials, electronics and waste.

Table 3.3. LCAs by subject matters / product groups (absolute numbers and %)

	CH	D	I	S	CH	D	I	S
	Absolute numbers				in %			
Motor vehicles et al / transport	11	46	15	23	7%	18%	27%	15%
Synthetic materials	5	44	6	6	3%	17%	10%	4%
Packaging	30	36	10	27	20%	14%	18%	17%
Hygiene/cleaning	0	23	0	5	0%	9%	0%	3%
Building / building materials	45	16	0	17	30%	6%	0%	11%
Food	9	11	0	6	6%	4%	0%	4%
Energy	31	11	1	9	21%	4%	2%	6%
Paper/printing	1	10	2	8	1%	4%	4%	5%
Electronics	3	8	7	13	2%	3%	13%	8%
Waste	8	8	5	12	5%	3%	9%	8%
Metals	3	5	0	3	2%	2%	0%	2%
Furniture	0	5	0	3	0%	2%	0%	2%
Textiles	0	4	2	0	0%	2%	4%	0%
Table wear	2	4	0	1	1%	2%	0%	1%
Others	1	12	2	17	1%	5%	4%	11%
Not known	0	17	5	7	0%	7%	9%	4%
Total	**149**	**260**	**55**	**157**	**100%**	**100%**	**100%**	**100%**

Like all the assertions made in this paper this empirical interpretation is subject to the proviso that only those studies could be considered that have been publicly disclosed. It is generally assumed that there is a high quota of strictly in-house LCA studies whose existence is simply unknown.

3.7 Conclusions

Looking back over the results listed in this chapter, it is clear that the inventory presented cannot give a complete picture of LCA activities in Germany, Italy, Sweden and Switzerland. Nevertheless, some conclusions can be drawn.

In general, the LCA activities have increased in the last years both in volume and intensity. This can be seen both in the number of companies involved in LCA activities and in the number of LCA studies published over the years. However, the picture differs among the four countries considered: there are three leaders and one laggard (Italy). Taking into account a normalisation by an indicator LCA-studies per ten billion GDP, Sweden and Switzerland seem to be more active in the area of LCA than Germany. The main users of LCA are business and industry. They produce the products which constitute the basis for the examinations within LCA-studies. In Switzerland, the role of the state as a client however is more important than in the other countries.

Looking at business, the application of LCA has also diffused through sectors and product groups. Whereas in the beginning, a lot of LCA-activities were conducted in some sectors for some specific issues (often packaging[9]), nowadays more sectors use this tool and also examine different product groups. However, most LCAs are made within larger companies, the diffusion into the SMEs world is still very modest.

For example, Swedish LCA activity is increasing according to the Barometer of 1995; more companies from more industrial sectors are now working with LCAs than in the previous Barometer of 1993. Although 'packaging' has been a typical application for LCAs, LCAs are now being made for a great range of products, such as building materials, car parts, white goods, and functions such as transportation, waste management and buildings.

[9] See Rubik/Teichert (1997).

4 A "static" perspective on LCA applications – Survey results

The application of LCA in business and industry is the topic of this chapter. The survey presented in chapter 4 does not consider a development-process, it provides us with a static and not a dynamic picture of the state of institutionalisation of LCA in different companies. The focus of the survey was an international comparison and we identified common and different patterns of LCA-application in Germany, Italy, Sweden and Switzerland.

Section 4.1 reports on the methodology, presents information on the sample, the return rates and method of evaluation. The following Section 4.2 reports on the different results[1]. Section 4.3 draws some conclusions.

4.1 Method, sample and evaluation procedure

4.1.1 Method

A survey of the application of LCA in four different European countries (namely Germany, Italy, Sweden and Switzerland) was carried out. A questionnaire with the same questions for each country was compiled; it consisted of five parts:

1. General information on the company,
2. The company and environmental matters,
3. Product innovation and the environment,
4. The use of LCA,
5. Future public environmental policy in the area.

Altogether, the questionnaire consisted of 35 questions. Nearly all of these were closed questions, many of them asked for rankings. The questionnaire was sent to the companies and these were asked for a written answer.

---

[1] However, this paper does not report on each result contained in the four national reports. It focuses on interesting findings and results; it considers differences between the four countries concerned. Therefore, interested readers are referred to the national reports for more details. The German (Rubik 1998a) and the Swedish report (Beckman/Baumann 1998) have been published, the other reports (Meier/von Däniken 1998, Mirulla 1998) not yet.

4.1.2 The sample

It was agreed to select approximately 400 different companies in each of the four countries. The companies were chosen according to two main criteria:

1. Environmentally oriented companies which were selected according to the following criteria:
   - *Switzerland*: Membership of ÖBU (Schweizerische Vereinigung für ökologisch bewußte Unternehmensführung) and companies which are certified according to ISO 14000 or EMAS.
   - *Germany*: Availability of environmental business reports displaying an outstanding quality; product-specific characteristics within the environmental reports; existence of an LCA-study; membership of German "green" industrial associations (UnternehmensGrün, B.A.U.M. – Bundesdeutscher Arbeitskreis umweltbewußtes Management, future); winners of environmental prizes or former clients of IÖW.
   - *Italy*: Selection from the database of Ambiente Italia.
   - *Sweden*: Membership of Swedish industrial organisations (Näringslivets Miljö Chefer), of ICC (Business Charter for Sustainable Development) or of Naturliga Steget.
2. Largest companies selected according to their turnover in 1996.

The mailing began in April 1997 and ended in May 1997. If possible, questionnaires were sent either to people within the companies known to the institutes carrying out the study or to environmental departments.

The sample and the response rates are presented in Table 4.1. A total of 1 625 companies received the questionnaire. 734 of these belong to the first group of environmentally-oriented companies; 891 to the second group of large companies. 382 usable and completed questionnaires were returned, a number corresponding to an average response rate of 23.5%. The proportion returned by the four countries differed considerably: In Italy, the return rate was 7.5%, in Germany 24.6%, in Switzerland 20.4 % and in Sweden 41%. However, these quotas corresponded to the expected return rate for each country due to the different specific national response "cultures".

19% of all answering companies are SMEs[2]. Especially in Switzerland, a lot of SMEs answered (quota: 41%); this is also due to the specific Swiss economic structure with a lot of relatively small and medium-sized companies.

[2] According to the criterion "Number of employees below 250". One also has to keep in mind that at least 50% of the companies in the sample were large companies according to the selection method.

Table 4.1. Sample and response rates

	Switzerland (CH)	Germany (D)	Italy (I)	Sweden (S)	Total / Mean
Total number of questionnaires	**403**	**410**	**400**	**412**	**1 625**
• to "environmentally" oriented companies	252	200	100	182	**734**
• to largest companies	151	210	300	230	**891**
Answers (absolute number)	**82**	**101**	**30**	**169**	**382**
Answers (in %)	20%	25%	8%	41%	**24%**
• from "environmentally" oriented companies	43	59	10	49	**161**
• from largest companies	18	45	6	72	**141**
• from "environmental" oriented and largest companies[a]	21	-	14	48	**83**
LCA users	44	62	18	66	**190**
LCA users' share of total number[b]	11%	15%	5%	16%	**12%**
LCA users' share of respondents	54%	61%	60%	39%	**50%**

a This classification has not been relevant in all of the countries.
b Of course, we do not know how many LCA users there are among the non-respondents.

4.1.3 Method of evaluation

As already mentioned, the questionnaire consisted of 35 questions. All of these were closed questions; the majority of these multiple-choice-questions offered several answer possibilities and allowed several answers; the number of allowed answers varied. Some questions offered rankings from "None" to "Crucial".

Companies not answering a question or a part of a question were generally treated as "refusal" – but only with regard to the specific question or sub-question. Refusals were excluded from any calculation, but they have been reported in order to assess their importance.

Questions involving a ranking were weighted in accordance with the following method: points allocated to the different answer possibilities (none = one point, low = two points, medium = three points, influential = four points, crucial = five points). Refusals were excluded. The weighted average values were calculated by means of the following formula:

$$\text{Average value} = (1*xi + 2*xj + 3*xk + 4*xl + 5*xm) / (xi + xj + xk + xl + xm)$$

i.e. refusals and "Don't know" answers were not taken into consideration in these calculations and calculated averages[3].

No analytical statistical evaluation was made; however, a descriptive statistic was carried out. For this purpose, the companies were separated into two groups according to a filtering question:

---

[3] However, the number of refusals is reported in the national reports.

- companies which declared their use of LCA and
- companies which declared they do ***not*** use LCA.

We rank and visualise the information in two ways: by figures, and by tables reporting a qualitative order by an A-B-C evaluation. The "A" refers to the three first positions of the ranking list (according to the weighted mean), while the last three positions of the same list correspond to a "C". All positions in between get a "B". Herewith, we obtained an impression of the relative national order and we tried to look for common (and/or) different application patterns in the different countries.

In a few cases we also compared the results of companies using LCA with the ones of those not using LCA. For this purpose, we used the results of the survey of LCA-using companies and divided these figures by the results of the companies not using LCA. The specific findings of this procedure are presented as (relative) percentages.

4.2 Results

In this section, we report on the different results of the survey. Subsection 4.2.1 reports on motivations to perform LCA-studies. Subsection 4.2.2 describes experiences of the application of LCA. Subsection 4.2.3 discusses the techniques of the application and the carrying-out of LCA-studies. Subsection 4.2.4 focuses on obstacles and the future. Subsection 4.2.5 contains an analysis of the relationship between LCA and product innovation. Subsection 4.2.6 is dedicated to policy implications and expectations.

4.2.1 Motivations for starting LCA

In this subsection, we report on motivations for starting LCAs within companies. Environmental concerns, the relationship between the application of LCA and environmental management systems, the importance of different stakeholders and the drivers for starting LCAs are described.

4.2.1.1 Environmental concerns and LCA

The companies were asked for their ranking of different environmental concerns[4], namely process-related concerns (water discharge, waste, air emission, noise, energy consumption), supplier-related concerns (environmental performance) and concerns related to the use and disposal of products. A ranking among five stages

[4] One possible answer for each concern was allowed.

was offered. The results were divided into companies using LCA and companies not using LCA:

- In *Switzerland*, LCA-applying companies have a stronger perception of environmental problems than companies not using LCA. Process-related and use/disposal-related concerns are ranked nearly identically.

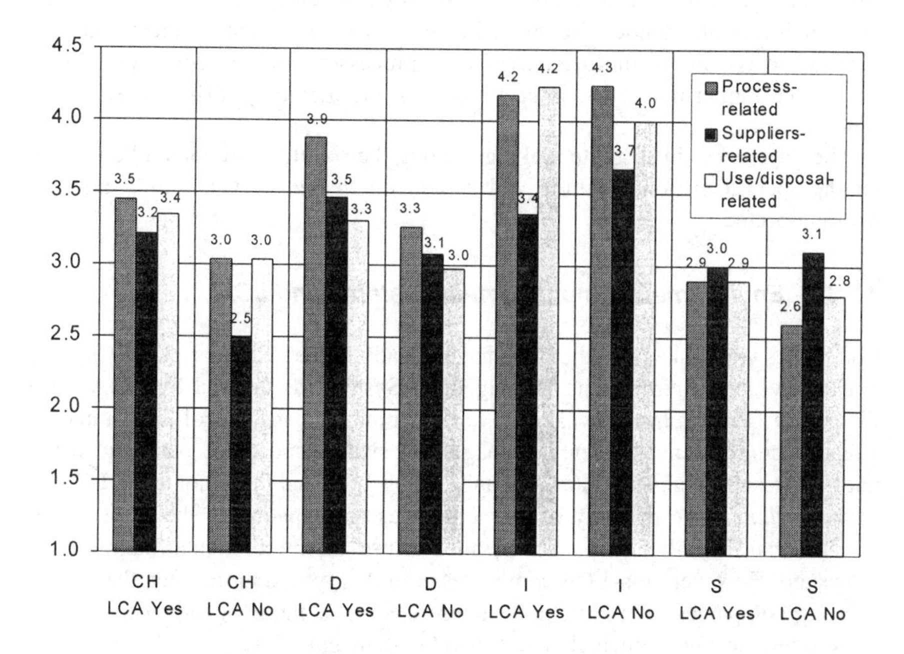

[One possible answer for each concern]

Fig. 4.1. Environmental concerns of LCA-using companies and those not using LCA - relative importance [*excluding* refusals]

- In *Germany*, LCA-applying companies have a stronger awareness of environmental problems than companies not using LCAs. In addition, there is usually a hierarchy in focus: first process-, then supplier- and finally use- and disposal-related concerns.
- In *Italy*, environmental concerns in nearly all cases are ranked higher than in the other countries. This may be an expression of the Italian "culture" of drama and action. In general, the differences between LCA users and non-LCA users are not important. One can conclude that environmental concerns have nothing to do with LCA. The process- and use/disposal-related environmental concerns are of the highest rank.
- In *Sweden*, LCA companies in general rank the importance of environmental concerns slightly higher than non-LCA companies. The largest difference is

that LCA companies rank their own process higher than non-LCA companies. It is not possible to tell if this is the reason for doing LCAs or if the awareness of the process importance come from the use of LCA. However, LCA-companies rank all three environmental concerns very similarly.

With the exception of the Italian results, LCA-companies – in general – have a stronger perception of environmental concerns than companies not using LCA. However, the difference between the two groups is quite small. From this fact, one might conclude that environmental consciousness seems to be a necessary but insufficient condition for LCA. Generally, process-related environmental issues are ranked higher than other issues; however, the hierarchy differs between the four countries.

The differences in absolute rankings among the countries are most likely based on the specific national culture and should be treated very carefully and not overinterpreted.

4.2.1.2 Environmental management systems and LCA

Companies were asked whether they have already implemented or plan to have or not to have an Environmental Management System (as defined by EMAS, BS 7750, ISO 14000)[5]. In the evaluation, we combined the answers of the companies already using an environmental management system and those planning to use one. The results were split into companies using LCA and those not using LCA:

- *Switzerland*: The majority of the companies participating in the survey indicated that they already used or planned to use an environmental management system. However, non-LCA companies used this system more often than LCA-using-companies. This might be interpreted as a more negative connection between the environmental management system and LCA.
- *Germany*: Most of the companies participating in the survey indicated that they had or planned to introduce an environmental management system. The differences between both groups are very small.
- *Italy*: The Italian situation is nearly the same as the German one; most companies participating in the survey indicated that they already had or planned to introduce, an environmental management system. The differences between both groups are very small.
- *Sweden*: Nearly all companies using LCA already had or planned to introduce an environmental management system. Here there seems to be a strong positive connection between environmental management systems and the application of LCA.

[5] One possible answer for each management system.

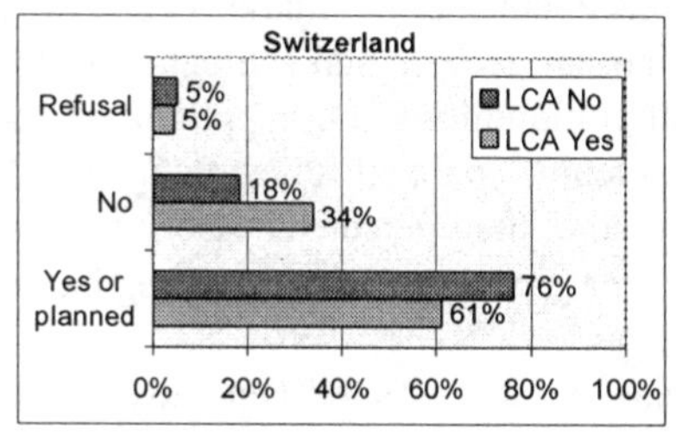

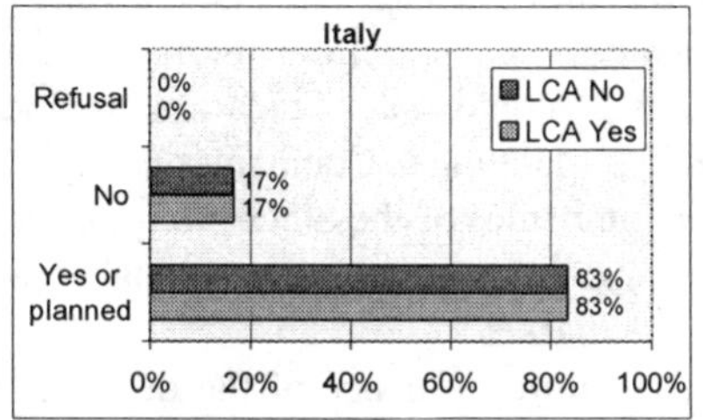

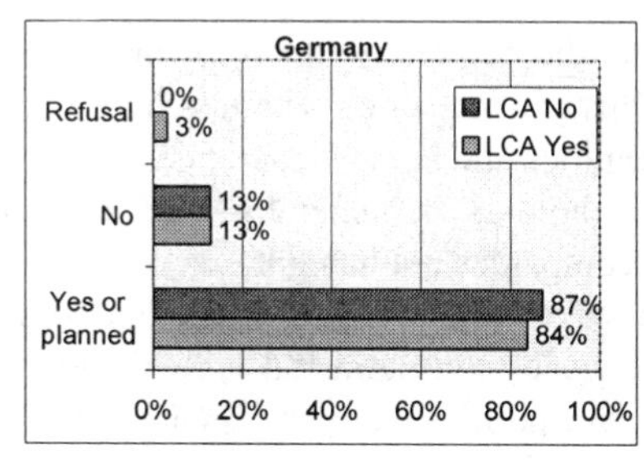

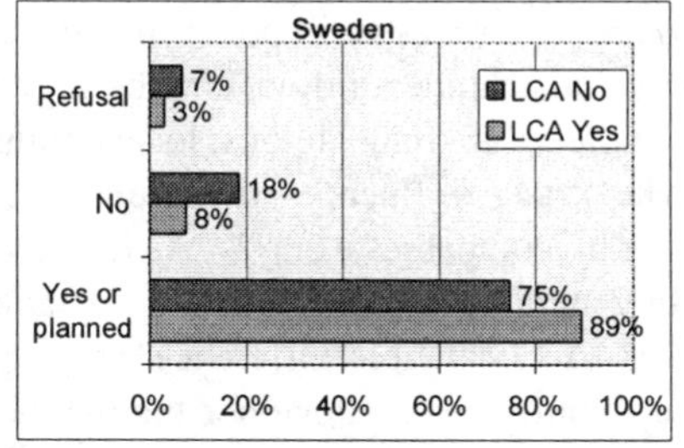

[One possible answer for each environmental management system]

Fig. 4.2. Environmental management systems and LCA (relative shares respectively in % of responding LCA-using and non-LCA using companies in each country)[6]

The general tendency seems to be that most responding companies are very active in the field of environmental management. Of course, this finding is also due to the selection criteria of companies adopted for the survey sample. However, the fact that a large percentage of companies not using LCA have or plan to have an EMS might lead to the conclusion that the existence of an environmental management system seems to be a necessary, but not sufficient condition for an LCA; in the case of Sweden, the positive relationship between the two seems to be stronger.

4.2.1.3 Importance of stakeholders and LCA

Stakeholders influence companies and their decisions and actions. We asked for a ranking of the importance of 11 different stakeholder groups according to a five step scale[7]. We also distinguished between the current importance and the expected future importance[8].

---

[6] It is worth remembering the numbers of LCA users as share of respondents in each country: CH: 54%, D: 61%; I: 60%; S: 39%.

[7] One answer for each stakeholder was allowed.

[8] Within five years.

- *Switzerland*: Business clients and final consumers (i.e. the market) as well as regulators (i.e. the policy) are the most important *current* stakeholders for both groups of companies. Trade unions, local communities and suppliers are the less important ones. Companies using LCA tend to rank all stakeholders higher than companies of the other group; in particular, they perceive a stronger pressure especially from environmental/consumer groups but also from regulators and the market.

  The future influence of the different stakeholders is rated generally higher than the current influence.
- *Germany*: Here, as well, policy-makers and the market are the most important *current* stakeholders influencing both groups of companies. The less important stakeholders are trade unions, local communities and banks/ insurances.

  The *future* influence of different stakeholders is generally also ranked higher by companies using LCAs than by companies not using it.

  In general, companies using LCA rank stakeholders higher than other companies. A remarkable difference exists between environmental and consumer groups and banks. Altogether the market and politics exert the greatest influence on both groups of companies.
- *Italy*[9]: The most important present stakeholders for both groups are the market (business clients and final consumers), followed by and the stockholders (i.e. owner, capital). However, the latter are ranked higher by companies not using LCA, while local communities are ranked higher by LCA-companies. Banks/insurances and NGOs are of a lower importance as stakeholders, but they are ranked significantly higher by LCA-using companies.

  For the future, nearly all stakeholders are ranked higher.
- *Sweden*: Market and politics are the most important current stakeholders for both groups of companies; however employees are considered to be an important stakeholder by companies indicating that they do not use LCA. LCA companies particularly perceive the pressure exerted by final consumers, environmental groups and media as more important than non-LCA companies. The non-LCA companies on the other hand rank local communities and stockholders higher than the LCA companies. In general, companies using LCA perceive external stakeholders stronger than companies which do not use LCA.

  The future influence of the various stakeholders is ranked in most cases very similarly to their present influence. However the importance of banks/insurances is supposed to become more important.

[9] Due to the specific national situation, the Italian questionnaire did not ask for media, suppliers and regulators as stakeholders. However, from general knowledge and from the case-studies carried out within the whole research project, it is known that at present these three groups are not the main motivation factors for starting LCA activities in Italian companies. The importance of regulators is expected to increase in the future.

The relative importance of the different stakeholders according to a ranking list is presented in Table 4.2 for those companies using LCA and in Table 4.3 for those companies which indicated that they do not use LCA. The general tendencies are:

- *LCA-users*: Final consumers, business clients and regulators are the most important stakeholders today and in the future. Trade unions are perceived as of no importance in this context. Environmental groups/consumer organisations, employees, stockholders, media are stakeholders of medium importance except in Switzerland where they are ranked higher in the future. No clear tendency exists for local communities, suppliers and a little surprisingly for banks/insurances.

Table 4.2. Relative importance of different stakeholders for companies ***using*** LCA in the four countries (relative ranking)[10]

Stakeholders	CH		D		I		S	
	today	future	today	future	today	future	today	future
• Final consumers	A	A	A	A	A	A	A	A
• Business clients	A	A	A	A	A	A	A	A
• Regulators	A	A	A	A	n.a.	n.a.	A	A
• Stockholders	B	B	B	B	A	A	B	B
• Environmental groups/ consumer associations	B	A	B	B	C	B	B	B
• Media	B	B	B	B	n.a.	n.a.	B	B
• Employees	B	B	B	B	B	B	B	B
• Local communities	C	C	C	C	A	A	B	C
• Banks, insurances	B	B	C	C	B	C	C	B
• Suppliers	C	C	B	C	n.a.	n.a.	C	C
• Trade unions	C	C	C	C	C	C	C	C

[One answer for each stakeholder]
A = high importance
B = medium importance
C = low importance
n.a. = not available

- *Non-LCA-users:* Regulators and, in most cases, final consumers and business clients are the most important stakeholders. However for Swedish companies employees play also an important role as present stakeholders. The role of stockholders in the four countries is different: in Switzerland and Italy, they are at the top of the ranking list whereas their importance is minor in Germany and Sweden. Media, employees, environmental groups/consumer associations are

[10] The method of ranking into A-B-C is described in subsection 4.1.3. As already mentioned, the Italian questionnaire did not ask for regulators, media and suppliers as stakeholders.

of medium importance. Trade unions are perceived as being of little importance – except in the Italian case.

Table 4.3. Relative importance of different stakeholders for companies ***not*** using LCA in the four countries (relative ranking)[11]

Stakeholders	CH		D		I		S	
	today	future	today	future	today	future	today	future
• Regulators	A	A	A	A	n.a.	n.a.	A	A
• Business clients	B	A	A	A	A	A	A	A
• Final consumers	A	A	A	A	B	A	B	A
• Stockholders	A	A	B	C	A	A	B	B
• Employees	B	B	B	B	B	B	A	A
• Media	B	C	B	B	n.a.	n.a.	B	C
• Suppliers	C	C	B	B	n.a.	n.a.	C	B
• Environmental groups/ consumer associations	B	C	B	B	C	B	B	C
• Local communities	C	C	C	C	C	B	B	B
• Trade unions	C	C	C	C	B	B	C	C
• Banks, insurances	B	C	C	C	C	C	C	C

[One answer for each stakeholder]
A = high importance
B = medium importance
C = low importance
n.a. = not available

Comparing LCA users and non-users (see Figures 4.3 and 4.4), it becomes clear that the tendencies are similar, but some differences do exist[12]:

- Present importance of stakeholders: German and Swiss LCA-using companies rank nearly all stakeholders higher than companies which do not use LCA; this is especially valid for environmental groups/consumer associations. Italian companies rank in a different way: similarly to companies in Germany and Switzerland, Italian companies using LCA rank environmental groups and local communities higher; however, stockholders, business clients and employees are ranked higher by non-LCA companies. Swedish companies also show a different picture; NGOs, the market and media are ranked higher by companies using LCA; politics (regulators and local communities), stockholders and banks/insurances are ranked lower.

[11] The method of ranking into A-B-C is described in subsection 4.1.3. As mentioned, the Italian questionnaire did not ask for regulators, media and suppliers as stakeholders. Therefore, we allocated the two major stakeholders to the "A"-group and the two minor stakeholders to the "C"-group.

[12] The method of constructing this index is described in subsection 4.1.3.

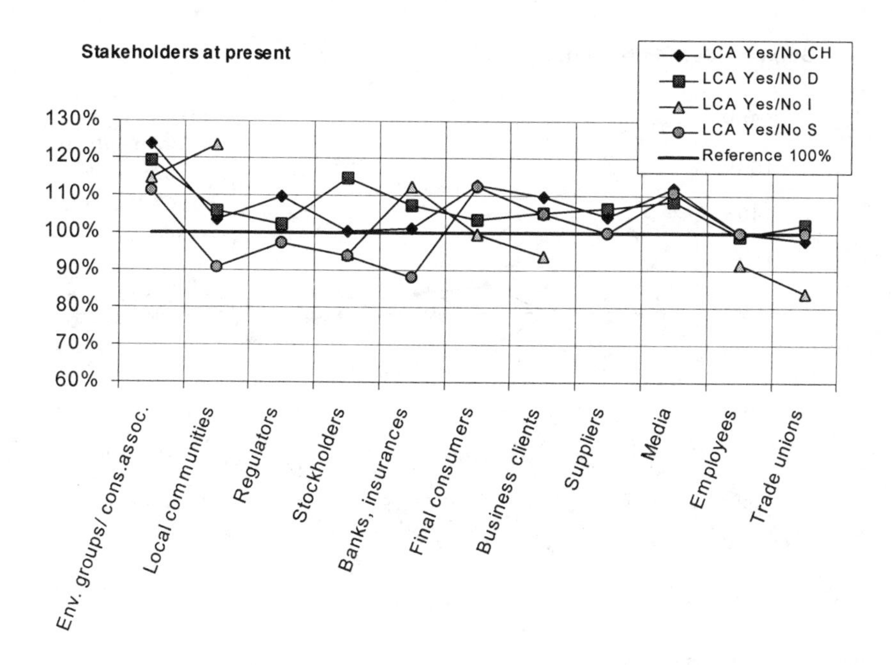

[One answer for each stakeholder]

Fig. 4.3. Differences in ranking of *present* stakeholders between LCA using and non-using companies (ranking weighted mean values of LCA companies divided by ranking weighted mean values of non-LCA companies – relative %)

- *Future importance of stakeholders*: Altogether, the future influence of the different stakeholders is rated higher than the current influence. LCA-companies tend to rank some stakeholders higher than companies not using LCA. As in the case of present importance, German companies using LCA rank nearly all stakeholders higher than companies not using LCA. Swiss companies rank the influence especially of environmental and consumer organisations higher; also media and regulators are ranked, to a modest degree, higher. The Italian picture is characterised by the huge difference for the ranking of local communities, which are of a much higher importance for LCA-using companies; for all the other stakeholders, the differences between both groups are not so important. The Swedish situation is also different because most stakeholders are not ranked higher by LCA-using companies.

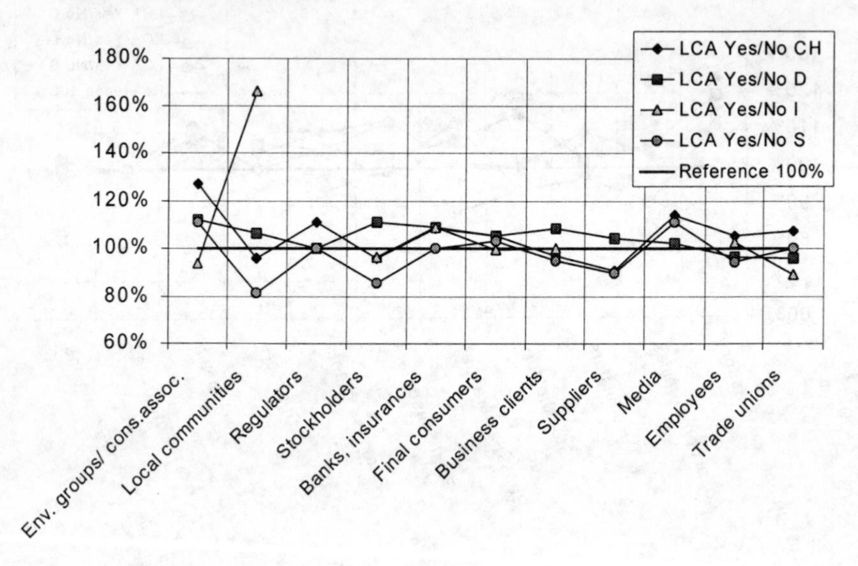

[One answer for each stakeholder]

Fig. 4.4. Differences in ranking of *future* stakeholders between LCA using and non-using companies (ranking weighted mean values of LCA companies divided by ranking weighted mean values of non-LCA companies – relative %)

4.2.1.4 Drivers for starting LCA

A lot of different impulses to start LCA exist[13]. This topic was covered by a specific question according to a five step scale for 13 possible drivers. Obviously, the results refer only to companies using LCA:

- *Switzerland*: Product-related environmental problems, cost-saving opportunities and emerging green markets are the most important drivers. The less important ones are encouragements from parent companies, use of LCA by competitors and the will to introduce new instruments for R&D.
- *Germany*: There are a lot of pushing factors which were ranked nearly identically: cost-saving opportunities, product-related environmental problems, emerging green markets, participation in collaborative LCA-studies, management decisions, perceived environmental discussions. Of less importance are the use of LCA by competitors, encouragements by the parent companies and the introduction of new instruments for R&D.

[13] Several answers to this question were allowed.

- *Italy*: The most important driver for starting LCA is the encouragement by parent companies. This is due to the structure of the companies answering the questionnaire (see subsection 4.1.2), which include a high quota of multinational companies. The next important drivers are cost-saving opportunities, new instruments for Research & Development (R&D) and perceived environmental discussions. The less important drivers are collaborative studies with external organisations, competitors who started to use LCA and initiatives by R&D.
- *Sweden*: Initiatives by R&D, product related environmental problems and cost avoidance are the most important drivers. Less important are the environmental legislation, the use of LCA by competitors and the use of LCA as a new instrument for R&D.

Table 4.4. Divers for starting LCA in the four countries[14]

Drivers	CH	D	I	S
• Product-related environmental problems	A	A	B	A
• Cost-saving opportunities	A	A	A	B
• Emerging green markets	A	A	B	B
• Decision of the management	B	A	B	B
• Perceived environmental discussions	B	A	B	B
• Cost avoidance due to future liabilities	B	B	B	A
• Collaborative study with external organisations	B	A	C	B
• Meet eco label criteria	B	B	B	B
• Initiatives by Research & Development	B	B	C	A
• Encouragement by the parent company	C	C	A	B
• New instruments for R & D	C	C	A	C
• Environmental legislation	B	B	B	C
• Competitors started to use it	C	C	C	C

[Several possible answers]
A = high importance
B = medium importance
C = low importance

One can conclude that the drivers are very similar in Germany and Switzerland: important are cost-savings and product-related questions (either as a perceived risk in the form of problems or as a chance to act proactively to respond to emerging green markets). The internal perception of external circumstances might be the most important driver to start LCA. Especially in Italy, the international context, expressed as dependence on external co-operation, is of huge importance. Worth mentioning is the most important Swedish driver "Initiative by R&D".

[14] The method of the ranking into A-B-C is described in subsection 4.1.3.

In all four countries, the direct influence of the application of LCA by competing companies is not perceived as a starting driver. Environmental legislation, i.e. political or legal pressure, is not an important driver for LCA, especially in Sweden.

4.2.1.5 Size and LCA

In the first place, it is worth considering the size of companies using LCA (see Figure 4.5). The subdivision is in line with the official Eurostat classification.

More than 1/3 of the Swiss companies indicating that they use LCA are SMEs, i.e. companies with less than 250 employees. This quota differs considerably from the quotas in Germany (14%), Italy (11%) and Sweden (9%). Large companies with more than 5 000 employees dominate in these three countries with quotas of more than 40%. Once again, the Swiss quota of 20% is considerably different.

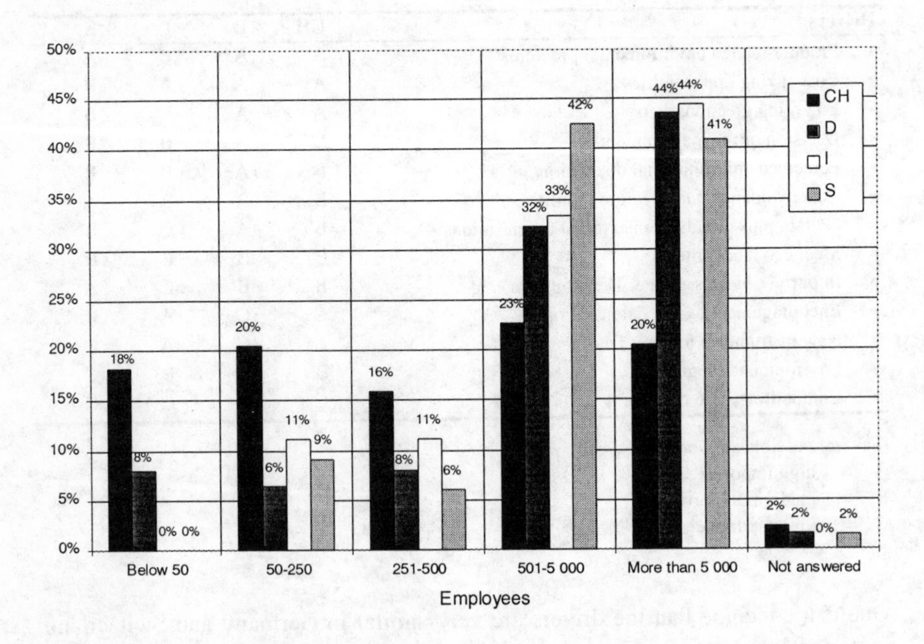

Fig. 4.5. Size of companies using LCA in the four countries

4.2.2 Application of LCA

The results of this subsection refer only to LCA-using companies and report on the different applications of LCA in business. They refer both to a set of different kinds of applications along the product development chain and to different kinds of products (some vs. all products, existing vs. new products, etc.)[15]. Both current and expected future applications are taken into account.

4.2.2.1 Current applications of LCA

Figure 4.6 shows the current applications of LCA in companies[16]. In order to compare the different countries, results are expressed in per cent.

Figure 4.6 and Table 4.5 (see subsection 4.2.2.2) highlight the main applications as well as the rarely used applications of LCA in the different countries. At present, the main results for each country are:

- *Switzerland*: The ranking of the most important LCA application is equally shared by the *identification of bottlenecks* and the *information and education of consumers and stakeholders* (57% of LCA-using companies declare to use it for these applications). The fact that 45% of companies say they apply LCA to *compare existing products and possible alternatives* is a (modest) hint towards the relationship between product innovation and the use of LCA in Switzerland. The less important applications in this country are two "strategic" applications, namely *radical changes in the product life cycle* and the *shift from product to service*.
- *Germany*: In Germany, the two most important applications are also the *identification of bottlenecks* and the *information and education of consumers and stakeholders* (61% and 60% respectively). The third ranking, at a much lower level however, is equally shared by *compare existing company products with products of competitors and procurement specifications* (32%). The application of LCA for *research development and design* is also very close to this ranking value (31%). Once again, the least important application are the *shift from product to service* and *radical changes in the product life cycle*. However, 15% of German LCA-companies say they use it for *radical changes in the product life cycle*. This might (modestly) suggest that German companies tend to use LCA in a slightly more "strategic" way than companies in other countries.

[15] For details about this possible classification of LCA applications, see subsection 2.2.2.
[16] Companies were asked to tick up to four choices.

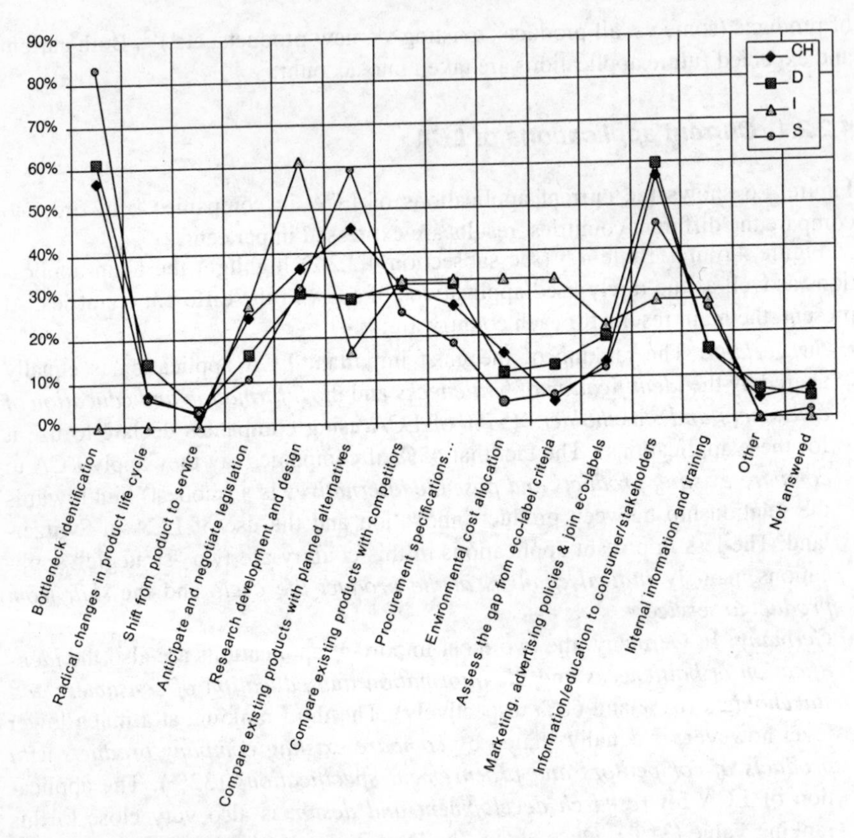

[Up to four choices possible]

Fig. 4.6. Current applications of LCA (relative preference shares in % of LCA-using companies in each country).

- *Italy*: Italy shows very different results[17]. LCA is mostly conceived for internal use in companies within the framework of research, development and design (R, D&D) activities (61%). This result is highly reliable. Whereas, a strange result is that no company said they used LCA for bottleneck identification. A possible explanation for this phenomenon is that the Italian term used in the questionnaire ("identificazione di colli di bottiglia") might not be common in the framework of Italian business and might not have been fully understood by the people filling in the questionnaire. Several other applications are used almost equally frequently by around 30% of companies. There is absolutely no

[17] "Absolute" Italian results, however, must be interpreted carefully, since only 18 companies responded saying they used LCA. Moreover, nine of these ticked only three choices, two companies two choices and one company only one choice.

sign of using LCA in a "strategic" way, as it is not used at all either for *radical changes in the product life cycle* or for *shift from product to service*.

- *Sweden*: The *identification of bottlenecks* is by far the most important application in Sweden (83%). There is also a rather clear tendency towards a more prospective use of LCA, since 59% of the companies use LCA to *compare existing products with possible alternatives*. LCA is also used for external information (*Information and education to consumers and stakeholders* – 45%). R, D&D follows with 31%. The less used applications are again *radical changes in the product life cycle* and *shift from product to service*, followed by *environmental cost allocation*.

Summarising, a quite common trend in Switzerland, Germany and Sweden exists, whereas Italy shows very different results. In the former three countries the identification of bottlenecks is the most important application of LCA (with a very high peak in Sweden). This application seems to be absolutely irrelevant in Italy[18]. Another common application in the three countries is for the external information of consumers and stakeholders. Once again, there is a big difference in Italy. This reflects the fact that in Italy people mostly think that LCA results are still too complicated to be communicated to the public. This might be connected to the fact that LCA in Italy is still at a rather early stage of development. As a matter of fact, a more "external" use of LCA results is expected in the future (see next paragraph).

The application for the purpose of comparing existing products and possible alternatives suggests a more proactive use of LCA in Switzerland and Sweden. This application is much less relevant in Germany (29%) and is rather rare in Italy (17%). On the other hand, Germany and Italy seem to be quite susceptible to pressure from outside (Compare existing company products with products of competitors; Procurement specifications...; Assess the gap from eco-label criteria, etc.). Finally, LCA is used in all countries as a tool for research, development & design. This is the main application in Italy, where LCA is still mostly regarded as an internal tool. However in the other countries as well, more than 30% of the companies use LCA for this particular application.

In all countries, LCA seems not to be used for the two "strategic" applications, i.e. *radical changes in the product life cycle* and *shift from product to service*. The only (modest) exception to this pattern is represented by the 15% of German LCA companies using LCA for *radical changes in the product life cycle.*

4.2.2.2 Future applications of LCA

Companies were also asked about expected applications of LCA within five years[19]. Figure 4.7 shows these expected future LCA applications. Once again, in

[18] However, as already mentioned, the question in the Italian version of the questionnaire could have been misinterpreted.

[19] Up to four answers to this question were allowed.

order to compare the different countries, results are expressed in per cent (of LCA-using companies in each country).

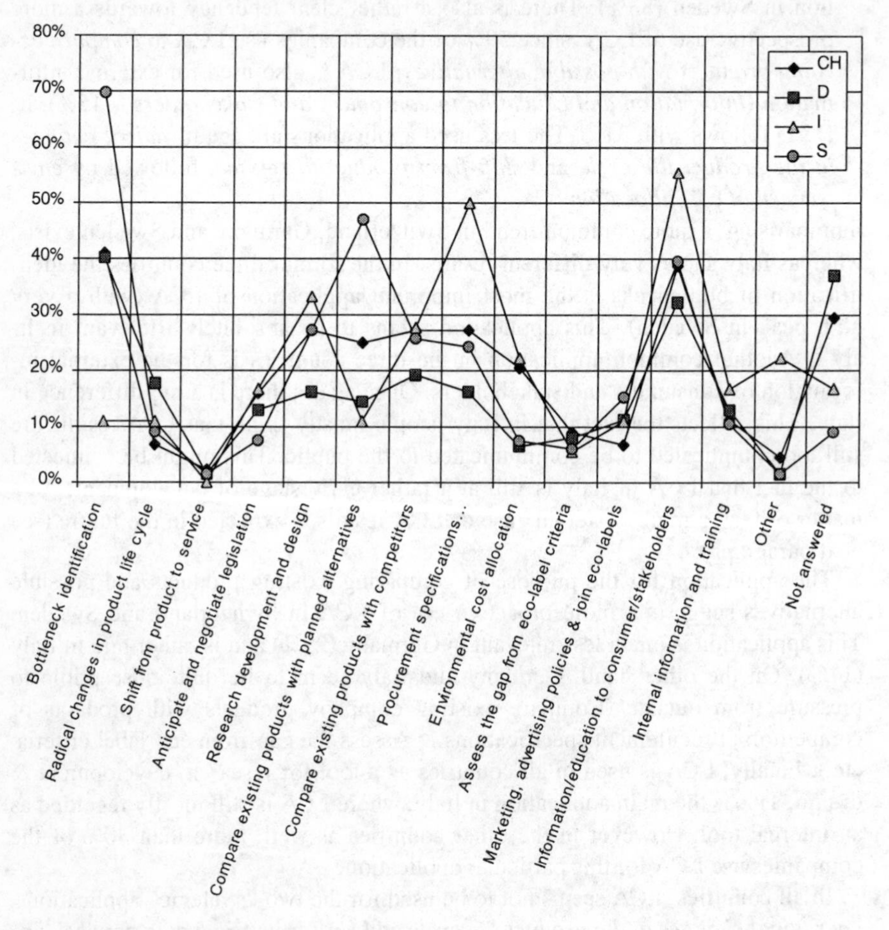

[Up to four choices possible]

Fig. 4.7. Expected future applications of LCA (relative preference shares in % of LCA-using companies in each country)

The following main observations can be made with respect to current applications. The three main application areas (the first two in Switzerland) remain the same in Switzerland, Germany, and Sweden. In Switzerland, the expected next most important LCA applications are *R, D&D*, *comparison with products of competitors* and for *procurement specifications* and *comparison with planned alternatives* . In Germany, the order of qualitative ranking is almost the same as at present (only *procurement specifications* lose importance). Percentage values are lower than today, but this is mainly due to a high rate of refusals[20].

In Italy there is a major shift towards more external applications, namely *information and education of consumers and stakeholders* (56%), and *procurement specifications, supplier screening, product co-makership* (50%). Also, the use of LCA as a more internal tool for R, D&D remains important (33%). In Sweden, there is practically no change between the choice of current vs. future applications. Only *anticipate and negotiate legislation and environmental cost allocation are ranked slightly lower than today.*

In all countries, there is a significant increase of non answering companies. Another interesting result is the low level of application for *marketing*, particularly if compared with the very high level of use for *information and education to consumers and stakeholders*. On one hand, this reflects the present high value of LCA for learning and for providing valuable shared information among different stakeholders; on the other it reflects the present difficulties and limits of its application for real marketing strategies (see also case-studies and conclusions).

Moreover, the application for *radical changes in the product life-cycle* shows a higher (although still modest) percentage than today. This gives a modest hint towards a more strategic use of LCA in the future. Finally, one must observe, that the application *assess the gap from eco-label criteria* is rated very low in every country. This might be explained by the fact that national eco-label procedures do not necessarily require an LCA, and that the European eco-label is still not known/applied.

Table 4.5 summarises both current and future application patterns for LCA in the different countries[21].

[20] This affects mainly absolute per cent values. However, the results in relative terms between different applications in each country are much more reliable.

[21] The method of the ranking into A-B-C is described in subsection 4.1.3. In addition, we used a threshold criterion: If more than 40% of all companies of a country indicated that a specific application is relevant, we ranked it with A; in case of less than 10%, we ranked it with C; all the other cases, we ranked with B.

Table 4.5. Qualitative ranking of present and future applications[22]

Application	CH		D		I		S	
	today	future	today	future	today	future	today	future
Bottleneck identification	A	A	A	A	C[23]	B	A	A
Information and education to consumers and stakeholders	A	A	A	B	B	A	A	A
Compare existing products with planned alternatives	A	B	B	B	B	B	A	A
Research development and design	B	B	B	B	A	B	B	B
Compare existing company products with products of competitors	B	B	B	B	B	B	B	B
Procurement specifications, supplier screening, product co-makership	B	B	B	B	B	A	B	B
Internal information and training	B	B	B	B	B	B	B	B
Anticipate and negotiate legislation	B	B	B	B	B	B	B	C
Marketing, advertising policies & joining eco-labelling criteria	B	C	B	B	B	B	B	B
Environmental cost allocation	B	B	B	C	B	B	C	C
Assess the gap from eco-label criteria	C	C	B	C	B	C	C	C
Radical changes in product life cycle	C	C	C	B	C	B	C	C
Shift from product to service	C	C	C	C	C	C	C	C
Other	4.5%	4.5%	6.5%	1.6%	0%	22%	0%	1.5%
Not answered	6.8%	29.5%	4.8%	37%	0%	17%	1.5%	9.1%

A = high importance (>40% of companies)
B = medium importance (between 10%and 40% of companies)
C = low importance (<10% of companies)

4.2.2.3 Kinds of products subject to LCA study

In order to identify whether companies use LCA in a rather retrospective or more prospective way, firms were asked which kinds of products they have analysed

[22] Future ranking values are generally lower, due to a much higher number of not answering companies, particularly in Germany and Switzerland.

[23] As already mentioned, this question might have been misunderstood in the Italian version.

with LCA[24]. Figure 4.8 shows the products which have been subjected to LCA studies in the various countries. Once again, in order to compare the different countries, the results are expressed in per cent.

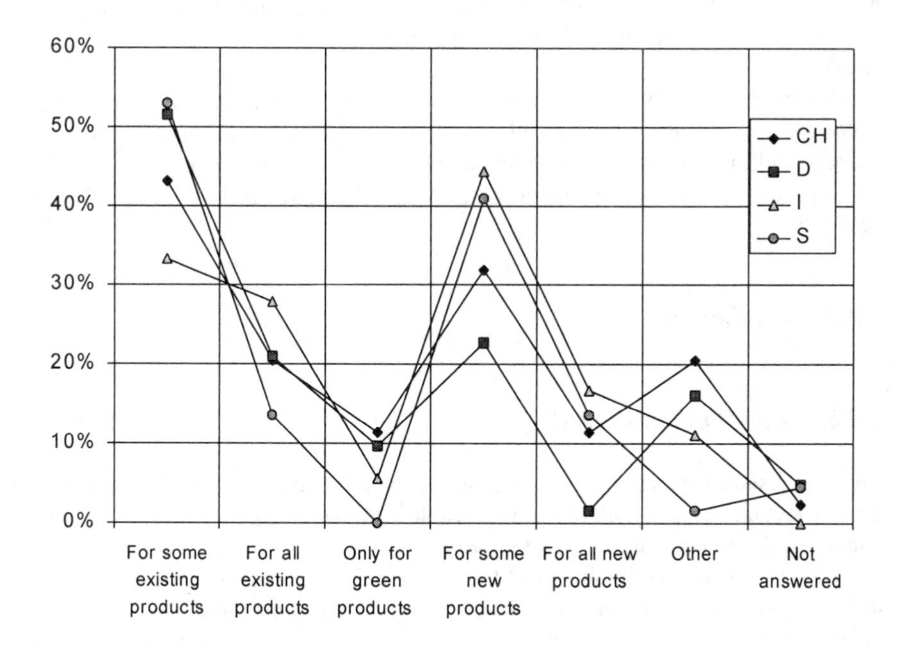

[Up to two answers possible]

Fig. 4.8. Products subject to LCA studies (relative shares in % of LCA-using companies in each country)

A common trend for all countries can easily be seen. LCA is generally applied to some products but not to all products. Moreover, LCA is mostly used for a few *existing* products, and is clearly *not* used for green products only. In general, at present LCA is still more frequently used in a retrospective way than in a prospective one, since it is applied more to existing products than to new ones.

In particular, the percentage of LCA application to all new products is low. All this suggests that LCA is not by any means used as a routine tool for product innovation, nor for environmental product innovation in particular. This seems to be especially true in Germany, where very few companies apply LCA to all new products, and 23% apply it to some new products. This seems slightly surprising, given the 29% of LCA-companies applying it for *comparison between existing products and possible alternatives* (29% - see subsection 4.2.2.1). In comparison,

[24] Up to two answers were allowed.

the higher percentage of Swiss and Swedish companies applying LCA to all new products and some new products is consistent with the higher rankings in subsection 4.2.2.1. This hints at a slightly more proactive use of LCA in those countries. The Italian results are not consistent with those of their previous ranking and do not seem to be highly reliable because of the low absolute number of respondents (all of the 18 LCA-companies responded, but nine identified only one preference).

Moreover, all these considerations have to be considered with some care, because the application of LCA to all or to new products also strongly depends on the size and branch sector of the company (i.e. a small company can have a little set of products, while a big chemical company might have thousands of different products).

4.2.3 Techniques

4.2.3.1 Functions involved

Figure 4.9 shows the company functions involved in LCA studies. The participation of several company functions is possible[25]. In order to compare the different countries, the results are expressed in per cent.

The set of functions involved in LCA is quite similar in all countries. The most involved officers/departments in all cases are the environmental departments (67-82%) and R&D (36-58%). Top management follows with the big exception of Sweden. This might be explained by the fact that Sweden is the country in which LCA is most developed and used, and by a different management "culture"[26]. As a matter of fact, the case-studies clearly show that in Sweden LCA is mostly introduced by a bottom-up approach.

Consistent with other results, in Italy LCA is used more within R&D activities and much less for marketing (see also subsections 4.2.2.1-3)[27]. On the other hand, health and safety officers are significantly more involved than in other countries. This is very likely caused by the organisational structure of Italian companies (see also subsection 4.2.5).

---

[25] Several answers to this question were allowed.

[26] The absolute numbers of companies answering the questionnaire and using LCA are quite similar in Sweden and Germany (respectively 66 and 62). However, if the difference of population and GDP is taken into account, it might well be concluded that Swedish companies are by far the most significant users of LCA among the four selected countries.

[27] By mistake, the purchasing department was not included in the Italian translation of the questionnaire. This result has therefore not to be taken into account.

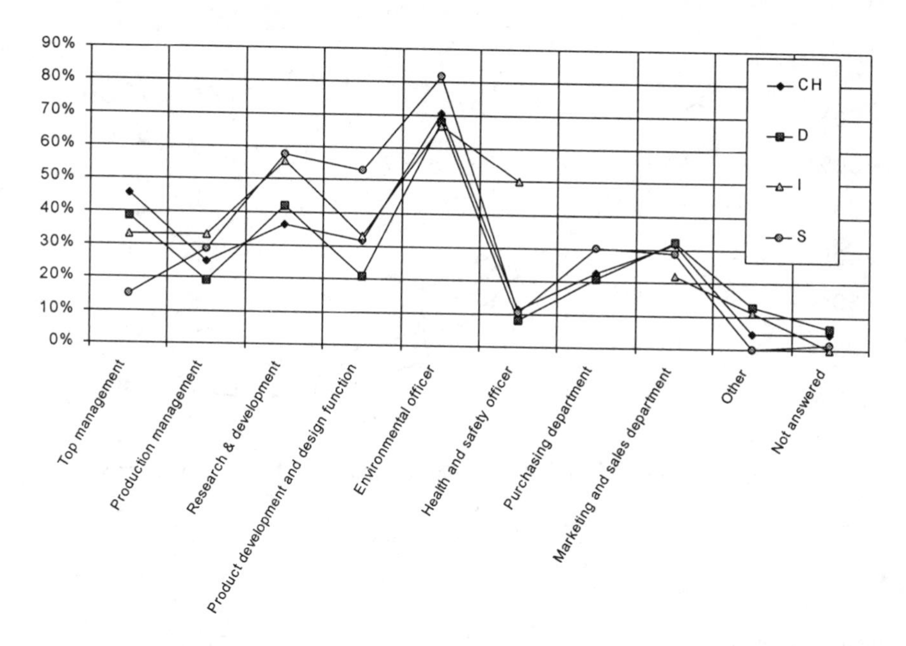

[Several answers possible]

Fig. 4.9. Functions involved in LCA (relative shares in % of LCA-using companies in each country)

4.2.3.2 Performers of LCAs

Companies were asked to identify the performers of LCA studies[28]. Figure 4.10 shows the percentage of LCA studies carried out by in-company teams vs. those performed by external consultants/research institutes or in collaboration with industrial associations.

[28] Several answers were allowed.

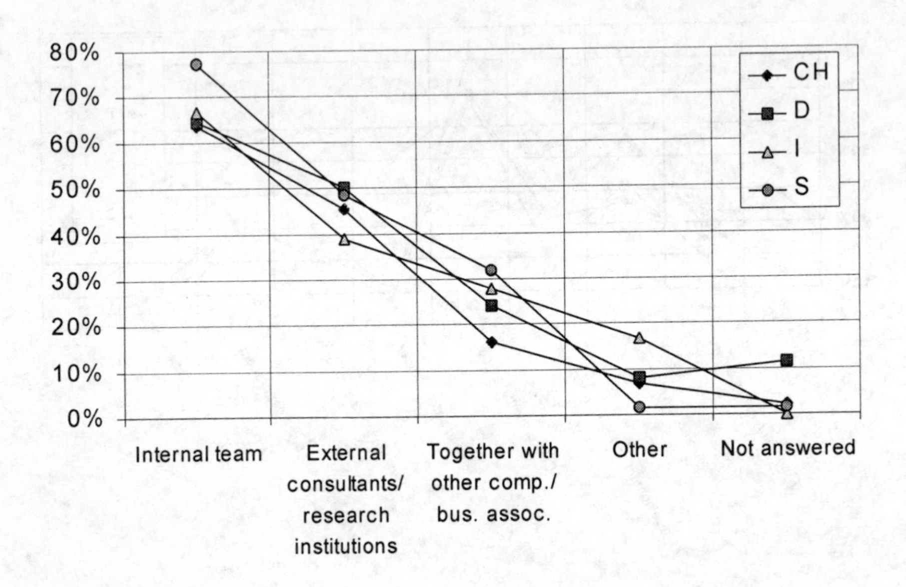

[Several answers possible]

Fig. 4.10. Performers of LCA studies (relative shares in % of LCA-using companies in each country)

In this case, the trend is very clear. In all countries, LCAs are carried out more and more by internal teams. In Sweden, the percentage amounts to 77% of LCA-companies. This suggests that the "internalisation" of LCA competences within the firm increases with the wider use of LCA. This result is also clearly confirmed by the case-studies (see chapter 5).

4.2.3.3 Problems

What are the main methodological difficulties companies have met in implementing LCAs?[29] Figure 4.11 shows the main methodological problems encountered by companies while performing an LCA study.

[29] Several answers to this question were allowed.

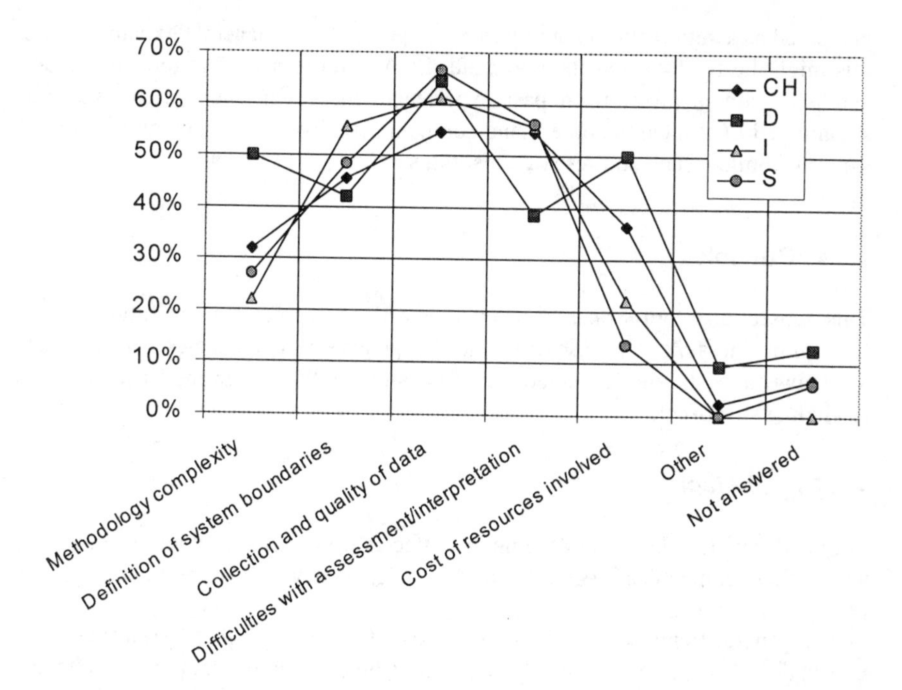

[Several answers possible]

Fig. 4.11. Methodological problems of doing an LCA (relative shares in % of LCA-using companies in each country)

Clearly, major difficulties are connected with the inventory step of an LCA. Data collection and quality are perceived as the biggest problem in Italy, Germany, and Sweden (from 61% to 67%). Switzerland (54%) seems to have a better data collection system. This might be explained by the existence of a variety of public data-bases carried out in collaboration with universities, research institutes and federal ministries and by the smaller size of the country. The second most important difficulty (40-50%), namely the definition of system boundaries, is also related to the inventory phase. This problem is particularly perceived in Italy (56%), where there is no great experience of LCA within companies.

As expected, a large group of companies has significant problems with the assessment and interpretation of results. This problem is perceived less in Germany (39% vs. around 55% in the other countries), but German companies refer much more (50%) to the general complexity of the methodology than other countries. The suggestion is that Italians are less expert in LCA and consequently less aware of the totality of problems. In Germany it might be an issue of "mentality", as frequently no interpretations and impact assessments are carried out. Curiously, only in Germany (50%) – and partially in Switzerland (36%) – there is a clear reference to the costs of the resources involved in LCA. This does not seem to be perceived as a significant difficulty in Sweden (13%). A plausible explanation for this might be twofold: On the one hand, LCA is more a routine tool in Sweden and practitioners profit from past experience; moreover, they often carry out streamlined LCAs. On the other hand companies in Sweden receive more external support from the state and research institutes.

4.2.4 Outlook

This subsection reports on the expectations of companies for the future use of LCA in business. A wider use of this tool depends on the obstacles, on the trade-off between costs and expected benefits, and on the experience accumulated (including surprises).

4.2.4.1 Obstacles

Companies were asked about the main obstacles preventing a wider use of LCAs within their companies[30] (see Figure 4.12) Per cent values refer to the sum of LCA-using companies.

The general trend is quite clear. Two results are similar in all countries: The first major obstacle to a wider use of LCA in business is the fact that results are disputable (40-60%). The second common result is the low ranking of difficulties to communicate results to top management. This result is quite surprising.

General methodological difficulties are perceived as a significant obstacle in Germany, Italy and Sweden. The Swiss "deviation" must be interpreted with some care, because Switzerland's rate of non-respondents in this case amounted to 20%. However, a plausible explanation for this might be that there is more public support from the state through its ministries and other important organisations (for example ÖBU, BUWAL) in this country.

The result referring to costs is consistent with the one of subsection 4.2.3.3: Costs are perceived as a main problem in Germany and Switzerland, but significantly less in Italy and Sweden. Interestingly, in no country is the cost of implementing measures suggested by LCA considered a main obstacle. To us, this is a rather surprising result. However this might be explained by the fact that many LCAs carried out until now have been retrospective/learning ones and not intended from the beginning as a design tool to introduce changes in production. Most of these studies have been carried out within environmental departments, which are "far away" from accounting and production departments. Another possible explanation is that the cost is accepted anyway before the study starts.

[30] Several answers to this question were allowed.

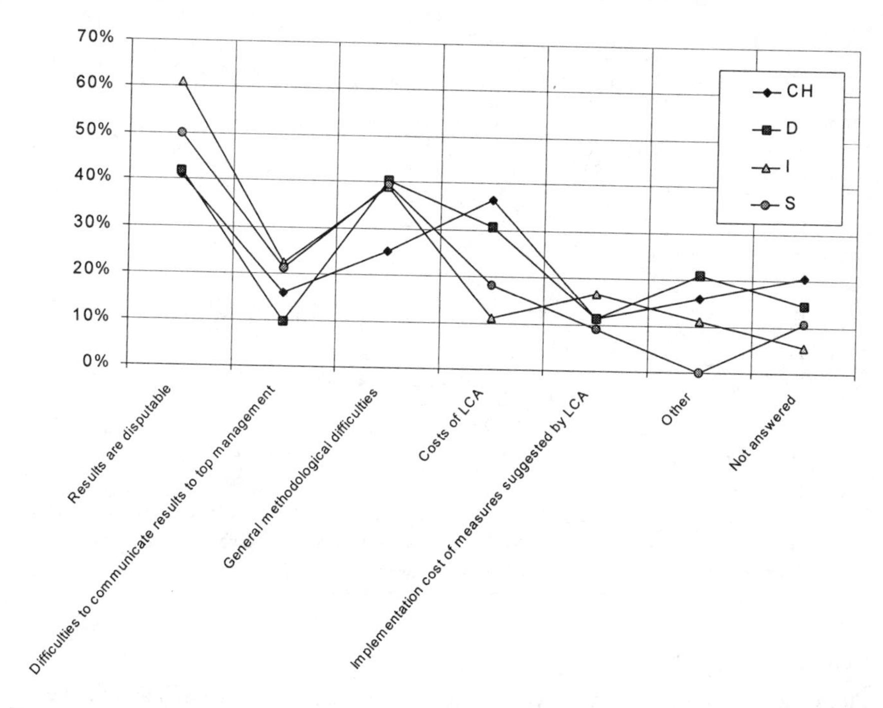

[Several answers possible]

Fig. 4.12. Main obstacles to a wider use of LCA (relative shares in % of LCA-using companies in each country)

4.2.4.2 Balance between costs and benefits

Companies were asked about the balance between costs and benefits of LCA[31]. Figure 4.13 shows how companies assess the benefits deriving from carrying out LCA.

There are three similar results in all countries and one big difference between two groups of countries.

[31] Several answers to this question were allowed.

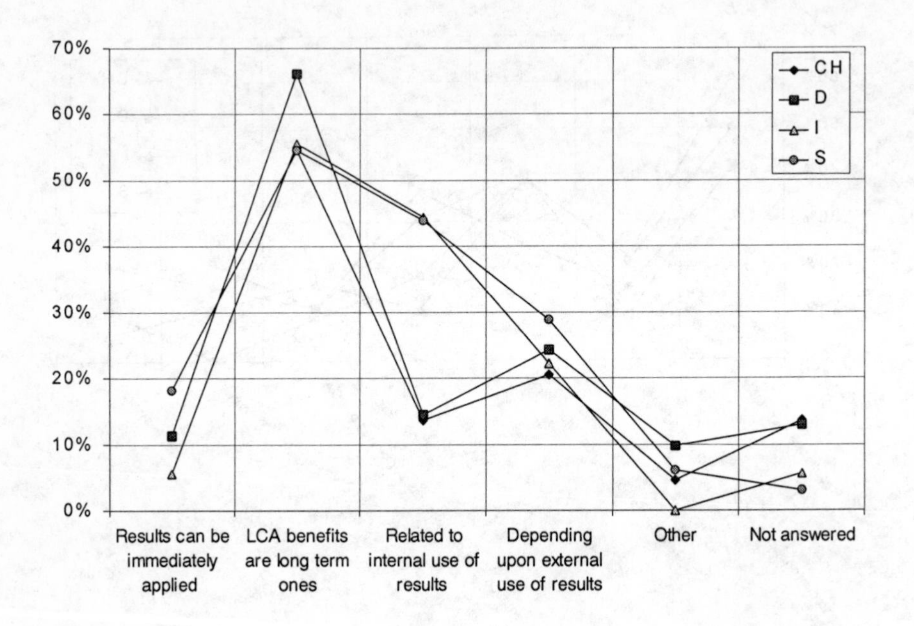

[Several answers possible]

Fig. 4.13. Benefits of LCA as they are perceived in companies in the different countries (relative shares in % of LCA-using companies in each country)

Companies in all countries do agree on the fact that results of LCA cannot easily be applied immediately and that benefits deriving from LCA are long-term. A quite large percentage (20-30%) of companies think that benefits depend on the possibility of diffusing results externally. However, results suggest that in Italy and Sweden LCA (and its benefits) is perceived mostly as an internal tool, whereas in Switzerland and Germany there is a stronger focus on the external use of LCA. This result is consistent, at least to some extent, with the results reported in subsections 4.2.2.1 and 4.2.2.2.

4.2.4.3 Surprises

Companies were asked if LCA produced any surprises[32]. Being asked whether they had had surprising results arising from the LCA, 44% of Italian companies answered *yes*, 28% *no*, and another 28% *do not know*. This result is consistent with the early stage of development of LCA methodology in this country. Germany has slightly more surprising results[33], while 47% of the Swedish companies

[32] Only one answer to this question was possible.
[33] However, this result may be influenced by the high rate of non-respondents.

answered *no*. Very clearly, Swiss companies were least surprised by LCA results (61% vs. 27%).

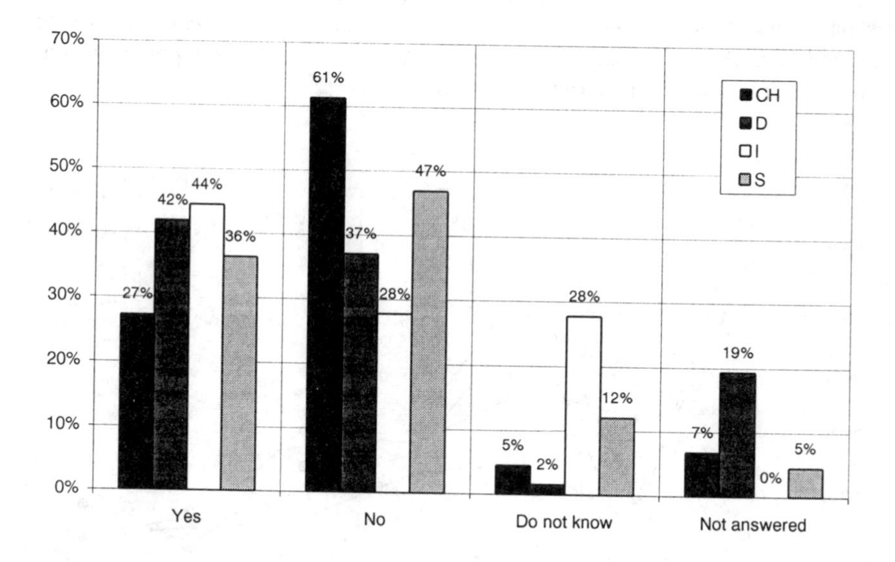

[One answer possible]

Fig. 4.14. Surprising results arising from LCA in different countries (relative shares in % of LCA-using companies in each country)

4.2.4.4 Increase of LCA-studies

In order to define an outlook for the future use of LCA in business decision-making processes, companies were finally asked if the use of LCA, in their opinion, would increase or not[34]. Figure 4.15 shows the answers from the different countries.

Companies are generally optimistic about the future use of LCA as a supporting tool for business. The main results can be summarised as follows: A large percentage of companies (up to almost 60% in Sweden) think that the use of LCA will increase in the future. Contrary to this, only a few companies (with the partial exception of 23% of Swiss LCA-using companies) think that the use of LCA will decline in the future.

[34] Several answers to this question were allowed.

The large difference between the answers to *only if used together with other instruments* suggests that Swedish companies are largely convinced that LCA will further develop on its own and does not necessarily need other auxiliary instruments to expand its role in business. This is further confirmed by the low response rate to *only if the methodology will be clearer* and *depending on the spread of LCA among companies*. This opinion is much less supported in other countries, particularly in Germany and Italy.

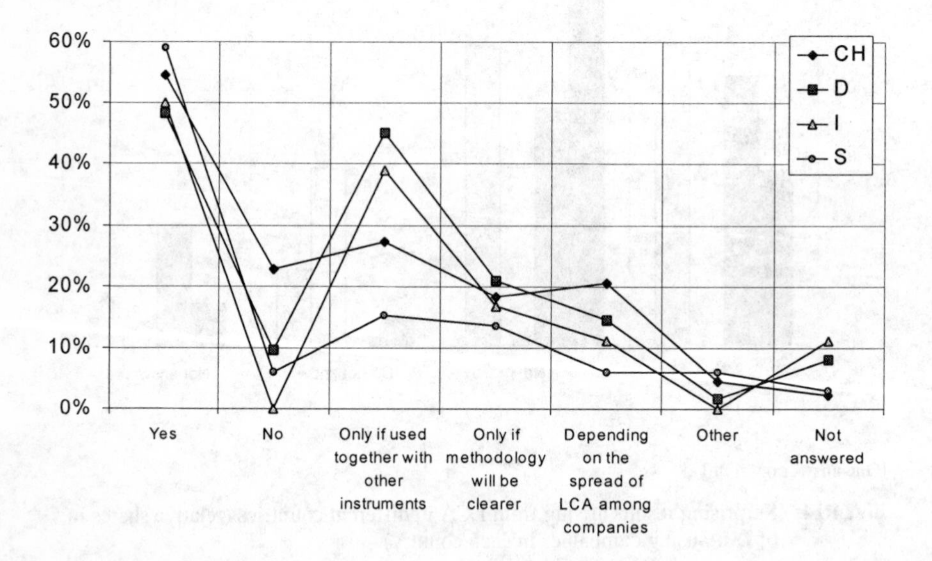

[Several answers possible]

Fig. 4.15. Increased use of LCA in companies in different countries (relative shares in % of LCA-using companies in each country)

4.2.5 Product innovation and LCA

This section tries to answer the question whether there is a connection between the use of LCA and (environmental) product innovation in companies. The first two sections show at which level of the company product innovation is defined and which the main drivers for change are. The next two sections analyse the functions in general and the extent to which environmental officers/departments in particular are involved in the process of (environmental) product innovation. The next subsection shows those management tools (including LCA) which are used most frequently in the context of environmental product innovation.

4.2.5.1 Definition level of product innovation

Figure 4.16 shows that product innovation is mainly defined at the following company levels (in order of importance): corporate strategy, marketing, and R&D[35].

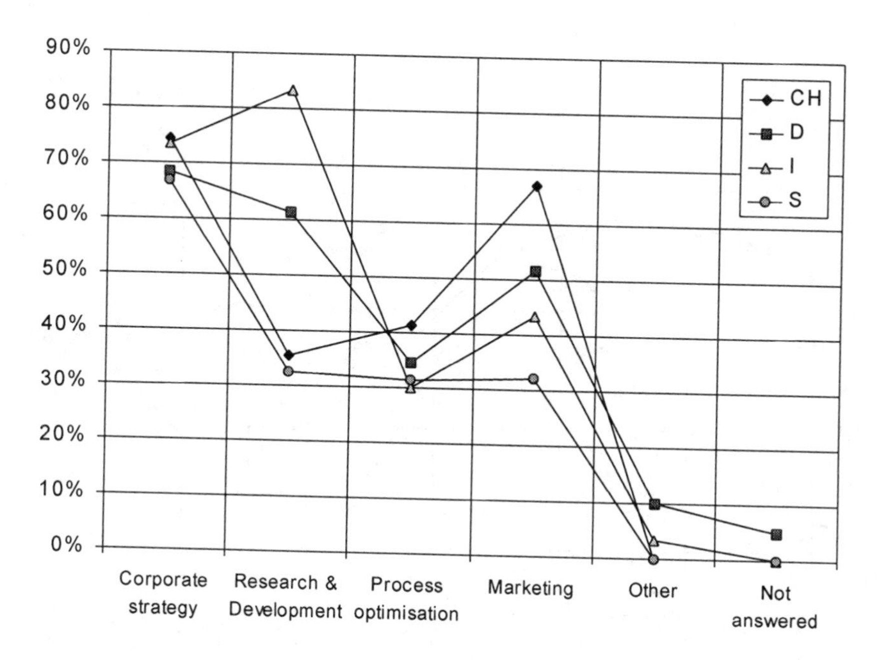

[Several answers possible]

Fig. 4.16. Company functions involved in the definition of product innovation (relative shares in % of all respondent companies in each country)

In the case of Sweden, product innovation is almost exclusively defined at the corporate strategy level. The Italian result regarding R&D must be handled with care, as it is neither consistent with the outcomes of the case-studies (see chapter 5) nor with other results (see subsection 4.2.5).

Results are expressed in per cent in order to allow comparisons between the countries. The percentage refers to the total of answering companies (LCA-using and non LCA-using ones). Several answers were possible.

[35] Several answers to this question were allowed.

4.2.5.2 Drivers for product innovation

Product innovations can be stimulated by different drivers. Figure 4.17 presents the ranking results of the main drivers for product innovation in the different countries[36].

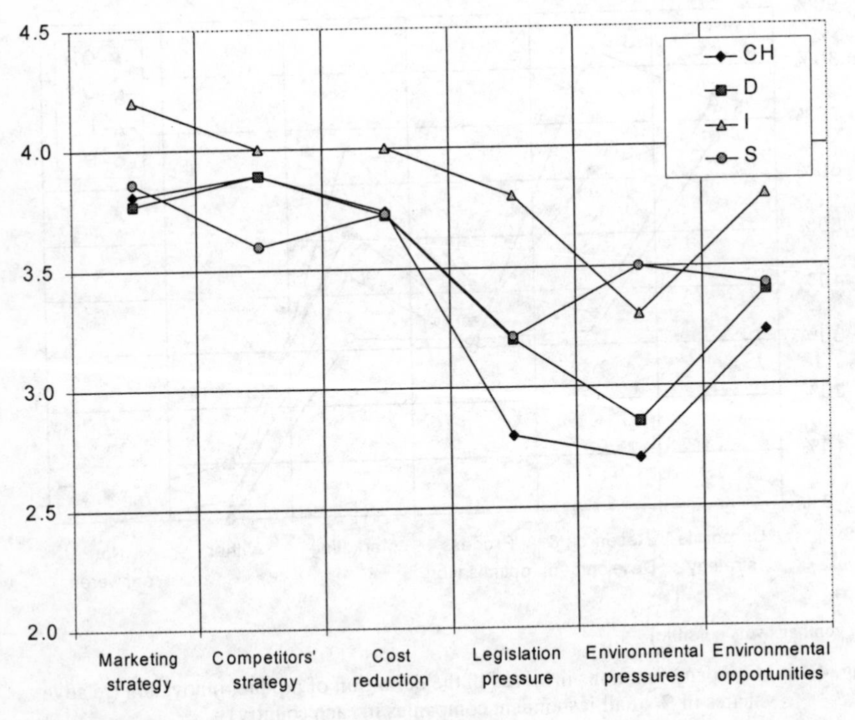

[Several answers possible]

Fig. 4.17. Drivers for product innovation (relative importance – weighted mean ranking values of all respondent companies in each country)

There is a quite similar trend in all countries. Firstly, product innovation is mainly driven by the market, i.e. costs and competition. There are only small differences between groups of countries as far as relative rankings of competitors with respect to marketing and costs are concerned. Secondly, everywhere environmental pressure is the least relevant driving factor for product innovation. Thirdly and interestingly, environmental *opportunities* are ranked higher than environmental and

[36] The method of weighting is described in subsection 4.1.3. Refusals are excluded from the calculation. A comparison revealed modest differences.

legal pressures. There is only a slight exception to this in Italy with respect to legal pressure. As already observed in other previous cases, rankings in Italy tend to be systematically higher than in other countries. This might reflect more a "cultural" emphasising attitude, than a real difference in evaluating the different driving factors.

Figure 4.18 shows the relative ranking value of LCA-using companies vs. non LCA-using companies. The indicator is constructed by dividing the average ranking of LCA-using companies by the average ranking of non LCA-using companies. The 100% reference case is when the rankings of the two groups of companies are equal.

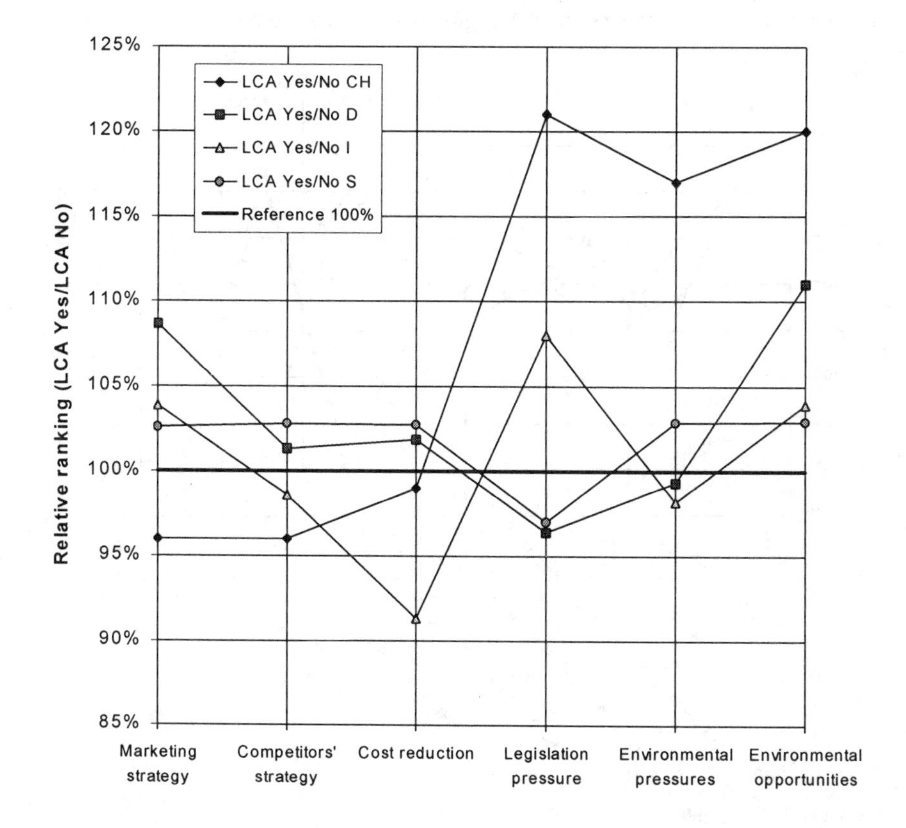

[Several answers possible]

Fig. 4.18. Ranking ratios of drivers for product innovation between LCA-using and non-LCA-using companies (ranking values of LCA companies / ranking values of non-LCA companies)

The situation is quite different in the various countries. In Switzerland, LCA companies rank *legislation pressure, environmental pressures* and *environmental opportunities* significantly higher than non-LCA companies. Swedish LCA companies rank all drivers slightly higher than non-LCA companies, with the only exception of *legislation pressure*. This situation is quite similar in Germany, where *marketing strategy* and *environmental opportunities* are ranked significantly higher. In Italy, it is worth mentioning the high ranking of *legislation pressure* and the low ranking of *cost reduction* by LCA using companies.

4.2.5.3 Involved departments / functions

We asked about the functions or departments normally involved in the definition of product innovation programmes[37]. Figure 4.19 shows the results.

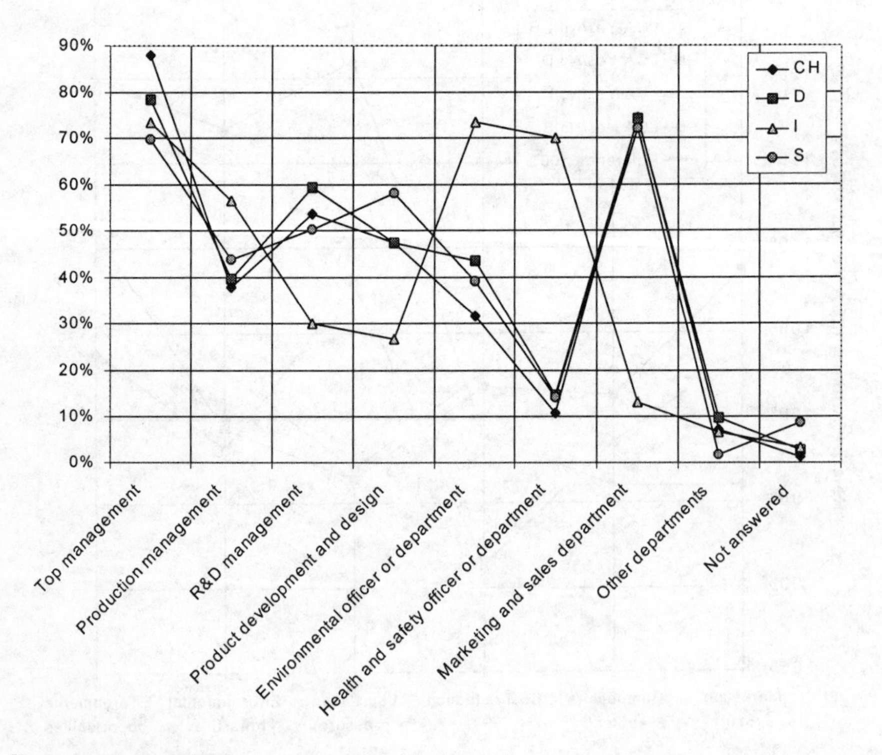

[Several answers possible]

Fig. 4.19. Departments/Functions involved in product innovation (relative shares in % of all respondent companies in each country)

In all countries, the most involved departments are the top management and the marketing and sales departments. The research, development, design and produc-

[37] Several answers to this question were allowed.

tion departments are much less involved (R&D and product development/design particularly less in Italy).

Italy shows a peculiar result: While in other countries environmental and health & safety officers are rarely involved in product innovation patterns (health & safety officers are least involved), 70% of Italian companies (the same percentage as for marketing departments) declare that their environmental and health & safety officers are involved in product innovation activities. A plausible explanation for this is twofold: On the one hand, in many cases the two positions coincide within the firm (this explains the registration of approximately the same value in the two cases). On the other hand, since the introduction in 1996 of a new law about safety, Italian companies have reacted by giving more importance and power to their health & safety departments.

4.2.5.4 Tools for environmental product innovation

Finally, the companies were asked which kinds of tools (including LCA) they use to improve their products environmentally[38]. Figure 4.20 shows the set of tools used by companies. Results are expressed in per cent to compare the situations in the different countries.

Figure 4.20 shows quite a similar trend in Switzerland, Germany and Sweden. Once again, Italy has some different results. In the former three countries, at least four tools (checklists, compliance/gap analysis with legislation, risk assessment and energy efficiency analysis) are used to more or less the same extent (30-50% of the companies). Material balances are used at the same level in Germany and Switzerland, but less so in Italy and Sweden. With the exception of Italy, environmental impact assessment is the least used tool for environmental product innovation. In fact, environmental impact assessment has been rather site-specific up to now and not related to products.

The other significantly different result for Italy is a much wider use of compliance/gap analysis with legislation. This might be interpreted as a rather reactive attitude of companies with respect to environmental issues.

Finally, an observation has to be made on the item *use of LCA*. The relatively low percentage shown in Figure 4.20 is due to including LCA-using and non-LCA companies in the calculation of the average percentage. The percentage of LCA-companies using LCA for making product environmental improvements, that is in a more prospective than retrospective manner, is of course significantly higher (61% in Switzerland, 53% in Germany, 67% in Italy, 29% in Sweden) than the value shown in Fig. 4.19. The fact that the Swedish percentage is the lowest is also due to the fact that 60% of the companies answering the questionnaire are *not* using LCA (see also subsection 4.1.2).

[38] Several answers to this question were allowed.

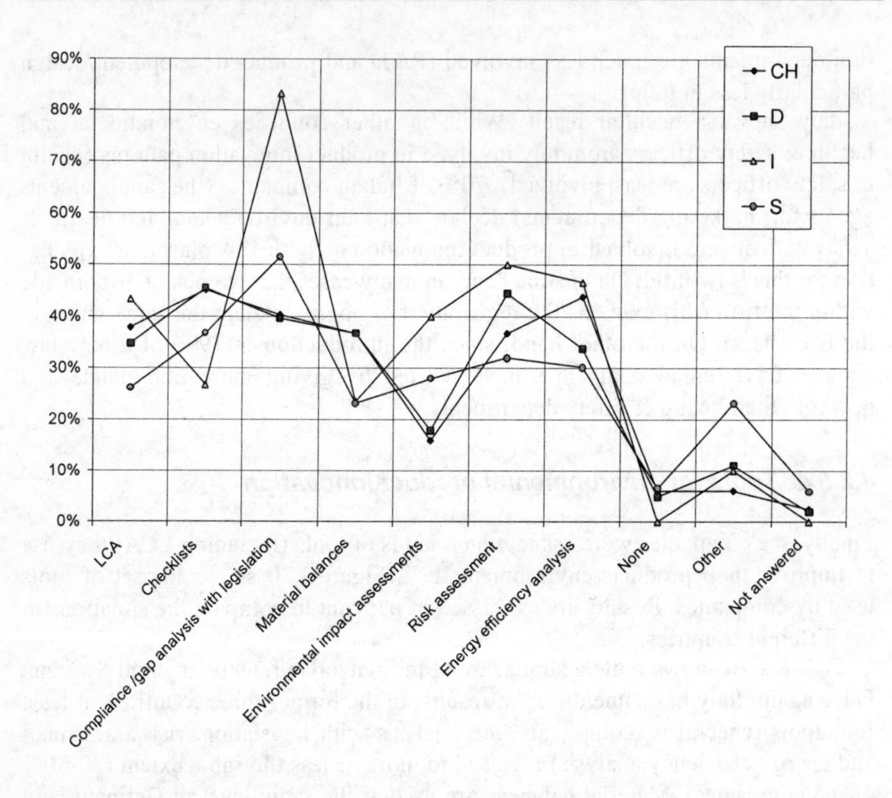

[Several answers possible]

Fig. 4.20. Mostly used management tools in the context of environmental improvements of products (relative shares in % of all respondent companies - both using LCA and not using LCA - in each country)

Curiously, also a few non-LCA companies *do* declare to *use* LCA or parts of it (inventory) to introduce environmental improvements into their products.

Figure 4.21 shows the differences between LCA and non-LCA companies, as far as the other tools are concerned (LCA itself is excluded for a better graphic scale). The parameter for comparison is the percentage of LCA companies divided by the percentage of non-LCA companies.

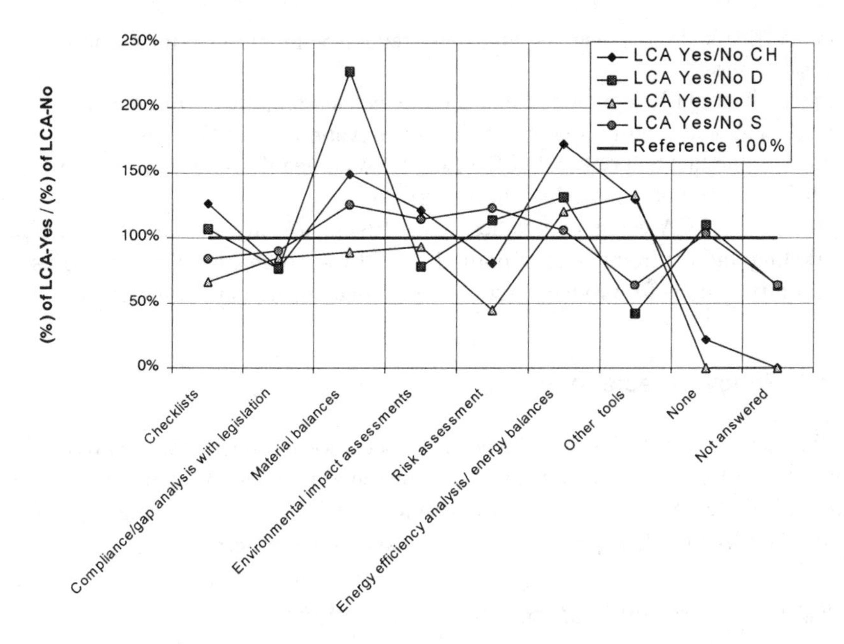

[Several answers possible]

Fig. 4.21. Differences between LCA and non-LCA companies as to the use of other tools (than LCA) for environmental product innovation (% of LCA companies divided by % of non-LCA companies)

The situation is a little different in the four countries. In Switzerland and Germany, LCA companies tend to use a larger set of tools for environmental product improvements.

- *Switzerland:* LCA-using companies apply more material and energy balances.
- *Germany:* LCA-using companies apply material balances much more frequently than non-LCA companies. In addition, they carry out energy efficiency analyses and/or balances as well as risk assessments more often. This may be related to the "history" of LCA, which was an evolution of energy and material analysis.
- *Italy:* LCA-using companies apply energy analyses and balances more than companies not using LCA. However, all other values are below 100%. This suggests that in Italy *non LCA* companies tend to use other tools for environmental product innovation more than LCA companies.
- *Sweden:* LCA-using companies apply material balances and risk assessments more than companies which do not use LCA.

These results suggest that, today, product innovation is driven by marketing, costs and competition. In all countries environmental pressure is the least relevant factor for pushing product innovation. The most involved departments are top management and marketing and sales departments. Environmental departments/officers do not usually take part regularly in product innovation processes[39].

Thus, it might be concluded that there is *not* a straightforward connection between LCA and (environmental) product innovation today. This is further confirmed by the fact that only 50-60% of the LCA companies said they applied LCA itself for environmental product improvements.

However, LCA-using companies rank drivers slightly higher, in particular marketing and environmental opportunities[40]. LCA companies also tend to use a larger mix of tools for product environmental assessment and improvement.

4.2.6 Policy expectations

In this subsection, we describe the policy expectations of business. We present the influences of current governmental environmental policies; prospects for the future and preferences of business for voluntary and/or mandatory measures. We also asked whether European or national measures are preferred.

4.2.6.1 Current political measures and actions

Governmental environmental policies may influence business activities in many ways. Therefore, we asked for a ranking of 13 different measures and activities which might affect business at present[41]. Once again, we distinguished between companies using and not using LCA. The results are:

- *Switzerland:* Covenants, product standards and certification schemes are the most influential present measures/actions according to both groups. LCA-companies tend to rank most of the measures higher than companies not using LCA; considerably higher rankings exist for public data bases for LCA (inventory of data) [+0.8 points] and for green design guidelines [+0.4 points].
- *Germany:* At present, the most influential measures for both groups of companies are *certification schemes* and *eco-auditing*; LCA-companies also rank *covenants* high whereas non-LCA companies rank *product standards* high. Comparing the results of both groups one recognises that some measures are ranked slightly differently by the companies, greater differences exist in the case of *covenants* [+0.5 points]; however, the general tendency is that LCA-companies tend to rank most of the measures higher than companies of the other group. Finally, it is important to observe the negative attitude of German companies with respect to LCA-based tax schemes.

[39] Italy seems to be an exception, but this may be due to the health & safety manager function, which often coincides with the environmental one in Italian companies.
[40] And legislation pressure in Switzerland and Italy.
[41] Several answers to this question were allowed.

Table 4.6. Correlation between different *present* political measures and the use of LCA in the four countries (ranking list)[42]

Action	CH		D		I		S	
	LCA yes	LCA no	LCA yes	LCA no	LCA yes	LCA no	LCA yes	LCA no
Certification schemes	A	A	A	A	A	A	A	A
Product standards	A	A	B	A	A	A	B	B
Eco-auditing	B	B	A	A	A	A	A	B
Product/packaging take-back systems	B	B	B	B	B	A	A	A
Covenants/sector codes of practice	A	A	A	B	C	B	B	A
Process standards	B	B	B	B	B	A	B	B
Eco-labelling	C	B	B	B	B	B	B	A
Green government purchasing	B	B	C	B	B	B	B	B
LCA-based tax schemes	B	B	C	C	B	C	n.a.	n.a.
Green design guidelines and awards	B	C	B	B	B	C	C	C
Green R&D programme	C	C	C	C	C	B	B	B
Green public investment funds	C	B	C	C	B	B	C	C
Sectoral LCI public data bases	B	C	B	C	C	C	C	C

[Several answers possible]

Explanation: A = high importance
B = medium importance
C = low importance
n.a. = not available

- *Italy*: At present, the actions which mostly affect both groups are *eco-auditing, certification schemes* and *product standards*. LCA-companies are also affected by *process standards*, non-LCA companies by *take-back systems*. Less affecting measures are sector LCI data-bases for both groups and covenants and green R&D-programmes by LCA-users. Comparing the results of both groups, it is a little bit surprising that companies which indicated that they do not use LCA rank most measures higher than companies using LCA.
- *Sweden*[43]: Swedish companies of both groups tend to rank *take-back systems* and *certification schemes* highest; LCA-companies state that *eco-auditing* affects them; companies not using LCA list *eco-labelling* and *covenants* at the top of the list. Nearly all actions are ranked higher by the companies using LCA in comparison to the companies of the second group; the most important higher rankings exist in the case of *eco-auditing* [+0.6 points], *certification*

[42] The method of the ranking into A-B-C is described in subsection 4.1.3.
[43] Due to the specific national situation, we did not ask about the tax-schemes.

schemes [+0.5 points], *public LCA data bases* [+0.6 points], *green design guidelines* [+0.5 points] and *green investment funds* [+0.5 points],

LCA *using* companies are especially affected by *certification schemes* (all countries), by *eco-auditing* (in all countries except Switzerland), by *product standards* (Switzerland and Germany), by *covenants* (Switzerland and Germany), and by *take-back systems* (Sweden).

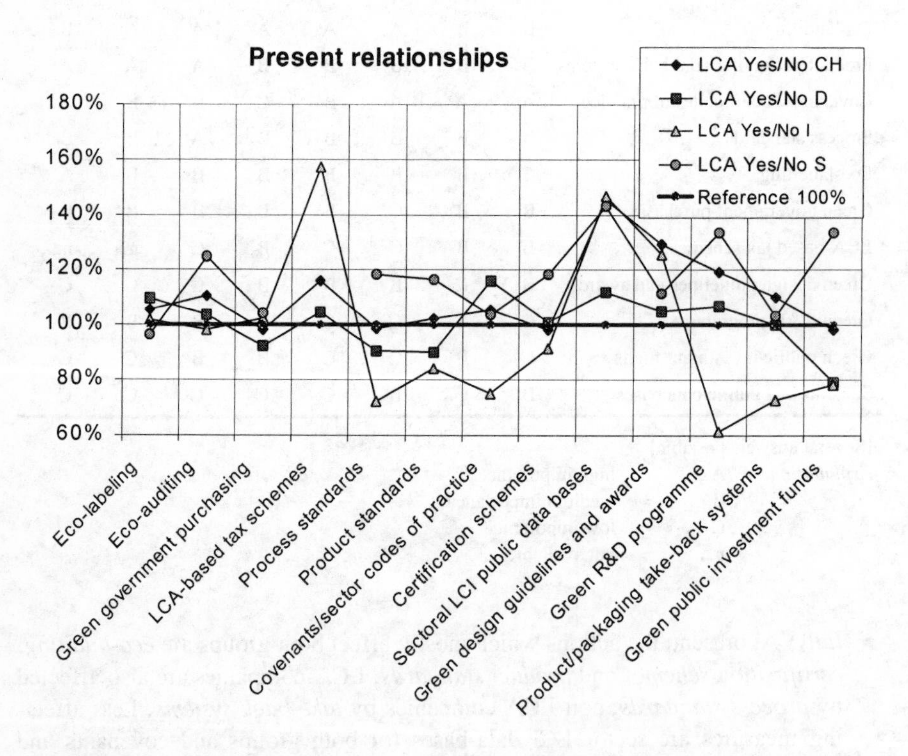

[Several answers possible]

Fig. 4.22. Relationship between companies using and not-using LCA with regard to *present* policy actions/measures [relative importances]

Companies which do *not use* LCA are especially affected by *certification schemes* (all countries), by *eco-auditing* (all except Switzerland), by *product standards* (all countries except Sweden), by *covenants* (all except Germany and Italy) and by *take-back systems* (all except Germany and Switzerland).

Summarising the main results according to this survey and the answers given by the participating companies, some present measures/actions affect them considerably, namely *certification schemes, product standards, covenants, eco-auditing and take-back systems*. At the bottom of this qualitative evaluation of affecting actions are *tax-schemes*[44], *public LCA-data, green design guidelines* and particularly *green R&D programmes*. Especially the role of *covenants* as a voluntary action seems to differ among the countries.

Are there differences in the perception of political measures/actions between the two groups of companies (see Figure 4.22)? In most cases, German and Swiss LCA using companies tend to believe that they are affected by policy actions/measures more than companies not using LCA. Swedish companies using LCA rank all policy actions/measures (except eco-labelling) higher than companies of the second group. However, the Italian situation differs from the three other countries; especially tax-schemes[45] and LCI-data-bases are ranked much higher by LCA companies; this is also valid for green design guidelines; in contrast to this, green R&D programmes, take-back systems, standards and covenants are ranked lower.

4.2.6.2 Future environmental policy measures

Governmental environmental politics may influence business activities in many ways. Apart from the current situation, we investigated expectations for the future by ranking measures and actions which might affect business within the next five years. The results are:

- *Switzerland:* The general tendency is that future measures are ranked higher than the present measures. The most important future actions which would affect both groups of companies are *certification schemes, covenants, product standards* and *tax schemes*; companies not using LCA indicated that also *take-back systems* and *eco-auditing* would affect them. The differences between both groups of companies are relatively small (less than 0.4 points).
- *Germany:* The arithmetical averages do not reveal great differences between both groups – with one exception: LCA-using companies rank *covenants/sector code of practice* an 0.5 point higher than the other group. The most influential actions are *eco-auditing* and *certification schemes* for both groups, *covenants* in the case of LCA-companies as well as *take-back systems* in the case of non LCA-companies.
- *Italy:* Similarly to the other countries, *certification schemes* are ranked at the top. LCA-using companies are affected also by *product standards* and *take-back-systems*; companies not using LCA will be affected in the future by *green public purchasing* and *process standards*. This is a hint that product and process standards are perceived differently by the two groups.

---

[44] However, one has to doubt that the respondents considered the relationship between eco-taxes and LCA-results; we suppose that they refer to eco-taxes in general and not to LCA based tax measures.

[45] However, one has to doubt that the respondents considered the relationship between eco-taxes and LCA-results; we suppose that they refer to eco-taxes in general and not to LCA based tax measures.

- *Sweden:* The tendency is the same as in the other countries. LCA-using companies rank future policy measures/actions on average 0.7 points higher and companies not using LCA by 0.8 points higher than present policy measures/actions. LCA-using companies rank especially *LCI-data-bases* higher in the future. Companies of the other group expect to be especially affected by *LCI-data-bases* and *eco-audit*.

The general trend is that future political actions/measures are expected to affect business more significantly than today.

Table 4.7. Correlation between different *future* political measures and the use of LCA in the four countries (qualitative ranking list)

Actions	CH		D		I		S	
	LCA yes	LCA no	LCA yes	LCA no	LCA yes	LCA no	LCA yes	LCA no
Certification schemes	A	A	A	A	A	A	A	A
Eco-auditing	B	A	A	A	B	B	A	A
Product/packaging take-back systems	B	A	B	A	A	B	A	A
Product standards	A	A	B	B	A	B	B	B
Covenants/sector codes of practice	A	A	A	B	B	B	B	B
LCA-based tax schemes	A	A	B	B	B	B	n.a.	n.a.
Green government purchasing	B	B	C	B	B	A	B	B
Process standards	B	B	B	B	B	A	C	B
Eco-labelling	B	C	B	B	B	B	B	B
Sectoral LCI public data bases	B	C	B	C	B	C	B	B
Green design guidelines and awards	C	C	B	B	C	B	C	C
Green R&D programme	C	C	C	C	C	C	B	C
Green public investment funds	C	B	C	C	C	C	C	C

[Several answers possible]

Explanation: A = high importance
B = medium importance
C = low importance
n.a. = not available

Are there differences in the assessment of future political actions between the two groups of companies (see Figure 4.23)? German, Swedish and Swiss LCA using companies tend to believe that they will be more affected by future political actions/measures than companies not using LCA. Especially eco-labelling, covenants, LCA data bases and green design guidelines (except Sweden) are expected to affect LCA-companies more. However, the Italian situation is different compared to the three other countries; in most Italian cases, companies using LCA rank political actions lower than companies of the second group.

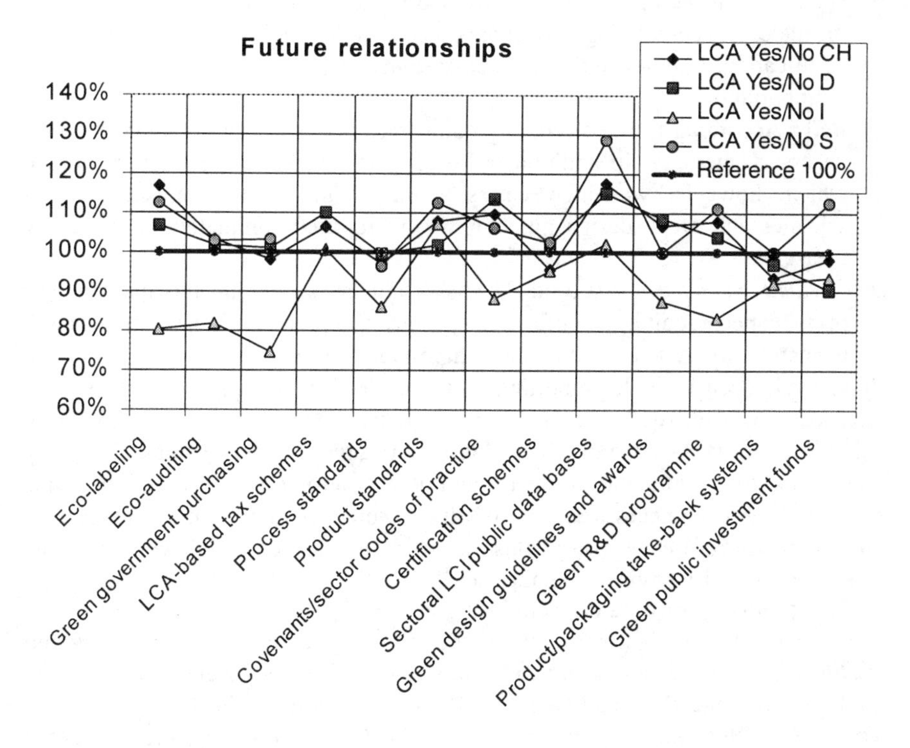

[Several answers possible]

Fig. 4.23. Relationship between companies using and not-using LCA with regard to *future* political actions/measures

4.2.6.3 Necessary actions and measures

In the two previous subsections, we asked companies if different measures affect their way of doing business; but: Which of the 14 different proposed measures *will be* necessary according to your opinion[46]?

We have not distinguished between both groups of companies, i.e. the following results fit the answers of both groups together:

[46] Several answers to this question were allowed.

- *Switzerland:* The companies answering this question declare that eco-auditing, covenants/sector codes, LCA-based tax-schemes, certification schemes and take-back-systems will be necessary.
- *Germany:* In the opinion of the responding companies five measures will be mandatory: eco-auditing, covenants/sector codes, LCA-based tax-schemes, certification schemes and take-back-systems. All other actions are relatively unimportant.
- *Italy:* According to the answering companies take-back-systems are the most necessary measure. In addition, a lot of other measures were listed, non of which, however, was assessed as similarly important. Interestingly, tax-schemes and green design guidelines were rejected completely as necessary measures.
- *Sweden:* Nearly every second Swedish company answering this question expects that eco-auditing, certification schemes and green R&D programmes will be mandatory. Also eco-labelling is important

In general, a considerable proportion of the companies returning the questionnaire did not answer this question. However, some tendencies become obvious: Once again, the answers of German and Swiss companies are very similar indicating that a mixture of voluntary measures and of very strong governmental measures are expected to be necessary. The Swedish companies think that especially voluntary measures are necessary; this statement has to be considered within the context of Swedish environmental policy[47].

The results as to political activities and measures assessed as necessary are very interesting: in all four countries, the voluntary actions of *eco-auditing* and *certification schemes* are particularly important. *Tax-schemes* and *covenants* are preferred in Switzerland and Germany, but rejected in Italy; this might be closely connected with national environmental policies. *Take-back systems* are considered as necessary especially in Italy, Germany and Switzerland; however, it is interesting to notice that Swedish companies assess this measure in a very different way. Some actions are considered as hardly necessary, e.g. *green design guidelines*.

[47] In Sweden, there are several different voluntary eco-labelling systems; environmentally oriented taxes and charges have been introduced (see OECD 1997).

4.3 Conclusions

In this section, we present general conclusions based on our findings of the survey.

4.3.1 Motivations for starting LCA

In general, environmental consciousness seems to be a necessary, but not sufficient condition for starting LCA. This conclusion is based on the observation that environmental concerns are of higher importance in LCA-using companies in two of the four countries, whereas in two countries concerns are ranked the same independently of the application of LCA. The existence of an environmental management system seems to be another supporting factor for carrying out LCA-activities. Especially Swedish companies show a deeper environmental involvement with regard to products and to the introduction of an environmental management system.

Business activities are influenced by a lot of different factors and groups. They are also influenced by stakeholders and the different business areas. Stakeholders are expected to have more influence in the future. Companies using LCA seem to be more influenced by stakeholders than companies which do not use LCA. One might conclude that LCA-companies tend to be more sensitive to external influences. The most important stakeholders are the market and public environmental policies; however, the influence of environmental groups and consumer organisations is perceived higher by companies applying LCA. This leads to the conclusion that external orientation and sensitivity might support the use of LCA. The role of environmental groups and consumer organisations must be taken into consideration; especially LCA-companies perceive the influences of these organisations. The Italian situation is slightly different because local communities are regarded as important stakeholders; this hints at a direct communication between companies and their surroundings[48]. In Sweden, some stakeholders are also less important for LCA-companies, namely public environmental policy, stockholders, banks/insurances and local communities.

Important drivers for LCA in all countries are cost savings; however, it is interesting to notice that the role of cost-savings as a driver is perceived differently in the four countries: They are mentioned explicitly as a driver in Germany, Italy and Switzerland, but in Sweden cost-savings are indirectly perceived via the future due to liabilities. This is a modest hint at Swedish companies being more proactively oriented than companies in the other countries. Other important drivers are product specific environmental discussions and problems. A specific Italian phenomenon is the huge influence of the international management of mother

[48] This conclusion might be taken with some care, because by mistake the Italian questionnaire did not include *regulators* and *media* and *suppliers* as stakeholders.

companies on the use of LCA. Another remarkable finding is the importance of R&D in Sweden which is another hint at the proactive orientation of Swedish companies. For all four countries, the direct influence of the application of LCA by competing companies is not perceived as a driver. Environmental legislation, i.e. political or legal pressures, is not important, especially in Sweden and Switzerland; however, in Germany environmental legislation is ranked close to the most important drivers. One might also conclude that a long-term and proactive orientation of companies supports the start of LCA because LCA is able to analyse and describe future problems and risks of products. Finally the relatively high importance of *emerging green market* as a driver for LCA in Germany and Switzerland (whereas in Italy and Sweden) has to be noticed. This is consistent with the results of the case-studies and reflects the (past) high *expectations* of Swiss and German companies about the use of LCA for marketing. In many cases these expectations could not be satisfied (see also next paragraph).

4.3.2 Application patterns of LCA

The analysis of *current* applications of LCA shows similar application patterns in Switzerland, Germany, and Sweden and different results in Italy. In the three former countries, the identification of bottlenecks is the most important application of LCA (with a very high peak in Sweden). Another common application in the three countries relates to the external informing available for consumers and stakeholders. In Italy on the other hand, LCA is mostly used as a rather internal tool for research, development & design activities[49]. This reflects the fact that *today* most Italian companies think that LCA results are still too complicated to be communicated to the public. This might be connected with the fact that LCA in Italy is still at rather an early stage of development. Anyway, in all countries the use of LCA for procurement specifications is ranked at medium level, while the application for marketing (final customers) purposes is low. This specific result is strongly confirmed by the case-studies. This reflects the fact that today, with the present state-of-the-art of the methodology, LCA can be used to inform (and influence) industrial clients and suppliers, but is still too complex (and often disputable) to be applied in the market of final consumers.

The application for the comparison of existing products and possible alternatives suggests a more proactive use of LCA in Switzerland and Sweden (see also subsection 4.2.2.3). This is also confirmed by the fact that these two countries have the highest share of companies applying LCA to all new products[50].

Common to all countries is that at present LCA is *not* used for the two "strategic" applications, namely for *shift from product to service* and *radical*

[49] However, in the other countries also more than 30% of the companies use LCA for this particular application.

[50] Italian results in this connection do not seem to be particularly reliable, since the answers to the two different questions are rather inconsistent with each other.

changes in the product life cycle (with the modest exception of Germany as far as the latter application is concerned). The application *assess the gap from eco-label criteria* is also rated very low in all countries. This might be explained by the fact that national eco-label procedures do not necessarily require an LCA, and that the EU eco-label is still not known/applied.

In *future,* only minor changes in application patterns are expected in Sweden (no changes at all in the preference order), Germany, and Switzerland. In Switzerland, increased application for *R&D*, *comparisons with products of competitors* and *procurement specifications* are expected.

On the other hand, in Italy a major shift towards more external applications, namely *information and education of consumers and stakeholders* and *procurement specifications, supplier screening, product co-makership* (but not *marketing*) is expected. However, LCA application of *R&D* is expected to remain significant.

The application for marketing purposes is expected to remain low everywhere[51]. In all countries, and particularly in Italy and Germany, the use of LCA for *radical changes in the product life-cycle* is expected to increase. This might suggest a (modest) hint at a *future* relationship between LCA and product innovation.

However, *today*, LCA is not yet applied as a routine tool for product innovation, and it is still used more in a retrospective than in a prospective way. This is suggested by the common trend in all countries indicating that LCA is mostly used for some existing products, and is clearly *not* used for green products only. In general, it is more frequently applied to existing products than to new products. Moreover, LCA is usually applied to some products and not to all products, demonstrated by the percentage of application of LCA to all new products which in Germany is particularly low. In this country, the application for *comparing existing products with planned alternatives* is much lower than in Sweden and Switzerland.

4.3.3 LCA-technique and outlook

In all countries the (most involved) governmental departments in LCA are respectively the environmental department, and the one of R&D. Top management follows, with the big exception of Sweden. This might be explained by the fact that Sweden is the country where LCA is most developed and used and/or by a different management culture. In Italy, health&safety officers are significantly more involved than in other countries[52].

In all countries, LCAs are more and more frequently carried out internally involving several departments/functions. In Sweden, this percentage amounts to 77% of all companies. This suggests that the "internalisation" and

[51] The expectiations refer to a time-horizon of five years.

[52] In Italy, a new law on health&safety has been established recently. Moreover, environmental officers often also cover the health&safety function at the same time.

"institutionalisation" of LCA competences within the company increases with a wider use of LCA[53].

Clearly, major difficulties are connected with the environmental inventory (collection and quality of data in Italy, Germany and Sweden, definition of boundary system in Italy). Switzerland seems to have a better data collection system, most likely connected with the higher availability of public data. As expected, a large fraction of companies has significant problems with the assessment and interpretation of results in all countries. Interestingly, in Sweden, the problem of costs is much less perceived than in other countries; and what is even more important is that LCA is more a routine tool. LCA users in Swedish companies profit from past experience and from more external support from the state and research institutes than in the other countries.

In general, most companies expect an increase in LCA-use for the future. Only few companies (with the exception of 23% of Swiss LCA-using companies) think that the use of LCA will diminish. Swedish companies are largely convinced that LCA will further develop on its own. In other countries, many companies expect that the use of LCA will increase, but in connection with other instruments.

The major obstacle to a wider use of LCA in business is the fact that results are disputable. General methodological difficulties are perceived as a significant obstacle in Germany, Italy and Sweden. Interestingly, in no country are the costs of the implementation of measures suggested by LCA considered a major obstacle. This might be connected with the fact that most of the LCAs carried out up to now were rather retrospective/learning LCAs and had not been intended from the beginning for implementation and environmental product innovation purposes[54].

As already mentioned, costs are perceived as a major problem in Germany and Switzerland, but significantly less so in Italy and Sweden. In all countries, most companies think that the results of LCAs are difficult to apply immediately and that LCA only brings about long-term benefits.

4.3.4 Product innovation and LCA

At present, product innovation is driven by marketing, costs and competition. In all countries, environmental pressure is the least relevant factor for pushing product innovation. The most involved departments are the top management and the marketing and sales departments. Environmental departments/officers do not usually take part regularly in product innovation processes[55]. Thus, it might be concluded that there is *not* a straightforward connection between LCA and (environmental) product innovation today. This is further confirmed by the fact

[53] This is strongly supported by the results of the case-studies as well.

[54] Another possible explanation is that the cost is accepted anyway before the study starts. This might particularly hold for Sweden.

[55] Italy seems to be an exception, but this might be due to the double function described in the previous footnote.

that only 50-60% of the LCA companies say they apply LCA for environmental product improvements. Instead, the majority of LCAs carried out up to now were rather retrospective/learning ones.

However, LCA-using companies tend to rank drivers slightly higher, in particular marketing (except Switzerland) and environmental opportunities (all countries). This suggests a correlation between a more strategic and proactive view of environmental product innovation and the use of LCA. LCA companies also tend to use a larger mix of tools for product environmental assessment and improvement (with the exception of Italy).

As already mentioned, some results presented above, suggest a tendency towards a more prospective use of LCA, which would obviously have major implications for the environmental innovation of products within companies.

4.3.5 Conclusions with regard to environmental policy and LCA

In general, business considers regulators as one of the most important stakeholders influencing companies. However, differences between companies using and not using LCA are not significant. A look into the future shows the same result. This means that the role and influence of politics on business is not expected to change.

Environmental legislation is not perceived to be among the most important drivers for starting LCA; nor do official eco-labels push LCA-activities significantly.

Political activities and measures affect business. The general tendency is that LCA-companies are affected more than companies not using LCA. At present, important business-affecting measures are certification schemes, covenants and eco-auditing as voluntary measures and product standards and take-back systems as mandatory measures. Especially eco-auditing, LCI-data bases and green design guidelines affect LCA-companies more than companies of the other type; these are hints that politically supported voluntary measures are able to stimulate the application of LCA.

The majority of companies tends to prefer voluntary instruments, namely eco-auditing and certification schemes. The other instruments are regarded differently in the companies and countries; clear preferences do not exist. Interestingly, neither public support of LCA by LCI-data bases nor standards (for both products and processes) are regarded as necessary measures. This is a clear signal for the demand for deregulation.

5 The dynamics of LCA adoption and integration in the firm – The results of the case-studies

In section 5.1, we introduce the applied method and give an overview of the case-study selection. The following section 5.2 presents twelve summary reports of our case-studies[1]. An overview of the application patterns of LCA based on all case-studies according to a common scheme is presented in section 5.3. This chapter closes with conclusions based on the case-studies (section 5.4).

5.1 Introduction

In addition to the quantitative survey approach, a qualitative approach was carried out. This in-depth-investigation dealt with the practices of the application of LCA in business and the role LCA plays in companies' decision-making. In order to do this, 20 case-studies in four countries (Germany, Italy, Sweden and Switzerland) were carried out, that is five in each country. Where multinational companies were concerned, their allegiance to a single country was not always lacking ambiguity (e.g. Kraft Jacobs Suchard: Formerly, Suchard was a Swiss company and although the access to the company took place through the Swiss branch, the case study actually concentrates on the German branch of Kraft Jacobs Suchard (KJS Germany)). A case-study is the examination of the application of LCA within one specific company. The 20 companies we looked at were selected according to the following criteria:

- existence of LCA-studies within a company,
- instructiveness (e.g. economic importance of company, LCA-pioneer),
- the willingness of a company to co-operate.

The companies derive from different sectors; in Table 5.1, we have classified them according to the NACE-categories elaborated by Eurostat (1996). As far as this is concerned, it has to be clarified that the allocation of companies to the NACE categories has been carried out according to the procedures of Eurostat, namely, the main area of economic activities. This is sometimes a little artificial, but it is the most appropriate way to present the companies.

[1] The summary reports present some selected information of the case-studies. Some longer and complete versions of the case-studies have been published separately (Bültmann/Rubik 1999).

Table 5.1. Overview of selected case-study-companies[2]

Sectors (according to NACE-terms)	Germany	Italy	Sweden	Switzerland
Extraction of crude petroleum and natural gas; service activities incidental to oil and gas extraction excluding surveying	–	–	Norsk Hydro	–
Manufacture of food products and beverages	–	–	–	Kraft Jakobs Suchard (KJS)
Manufacture of pulp and paper and paper products	–	Cartiera Favini	–	–
Manufacture of chemicals and chemical products	Henkel Weleda	GlaxoWellcome Italy	Akzo Nobel Perstorp Flooring	Ciba Speciality Chemicals
Manufacture of other non-metallic mineral products	Ytong AG	–	–	–
Manufacture of fabricated metal products, except machinery and equipment	–	–	–	Ernst Schweizer
Manufacture of machinery and equipment n.E.C.	AEG Hausgeräte Bosch and Siemens Hausgeräte (BSH)		–	–
Manufacture of electrical machinery and apparatus n.E.C.	–	ABB Italy	–	Landis & Gyr
Manufacture of radio, television and communication equipment and apparatus	–	Italtel	Ericsson	–
Manufacture of motor vehicles, trailers and semi-trailers	–	Fiat Auto	Volvo	–
Construction	–	–	–	Holderbank Cement and Beton (HCB)

Table 5.1 shows that there was a main focus on the chemical industry with six companies. In addition to that, we investigated companies from ten other sectors.

In order to structure and compare the collected information it was decided to carry out all case studies in a, as far as possible, systematic and symmetric way; this meant that the procedure concerning method, structure and interviewed persons and institutions were almost standardised. Therefore, we developed a

[2] More descriptive information on the case-studies can be found in the overview tables (see section 5.3).

"guidebook" with a list of research topics which were to be examined during the research process, namely

- the company and its background,
- description/characterisation of the LCA-study,
- description of business-external context (legal and socio-economic context),
- motives / background of an LCA,
- decision-making "culture",
- industrial and country-specific "culture",
- objectives of the LCA,
- organisation of the LCA,
- interaction and communication,
- decision, information,
- assessment of the case-study,
- future policy relevance of LCA.

Each of these twelve points contained several specific questions. These topics and the questions were to be used during the work on each case-study.

Data and information for the case-studies were collected in different ways, the most important were:

- interviews (verbal and by phone),
- observation of meetings,
- analysis of written documents.

Although the interviews followed the agreed determined line of questions they had predominantly a narrative character. In some case-studies, diaries were kept.

The collected information has been brought together, analysed and presented in a case-study report. A unique structure for case-study reporting was agreed upon; this structure encompassed six main parts, namely:

- introduction and background information,
- the selected LCA-study,
- role of LCA in the decision making context,
- assessment and evaluation,
- conclusions,
- references and annexes.

For each of the twenty case-studies, a complete and long case-study report was prepared[3].

Before presenting the summary of the case-studies, there is a short introduction about the national context of the selected case-studies for each country.

[3] In one case, the company did not accept to use and disseminate the collected information.

5.2 Summary reports of selected case-studies

The documented information on the 20 case-studies which we carried out was too large to present here in its entirety, therefore this subsection we present only summary reports of the more important case-studies. We restricted ourselves to presenting summary reports of 12 case-studies, each one following the same structure, namely:

1. The company,
2. The LCA-study which was examined by us,
3. Present and future use of LCA in general.

Before presenting the companies we illuminate the national situations in the four countries considered.

5.2.1 Switzerland

5.2.1.1 National situation

In Switzerland, LCA was introduced very early, mainly by a study on packaging, which was published by Swiss authorities in 1984 (BUS 1984). During the following years many studies were carried out by universities. In the beginning, the tool was hardly used by companies, because it had very restricted areas of application and few data were available. In 1990, a new tool for the interpretation/weighting stage was developed (cp. Ahbe/Braunschweig/Müller-Wenk 1990). Since then it has been used by an increasing number of companies. This screening tool allowed simplified decision making in purchasing or development of products. During the last years, it has been used more and more in the context of environmental management systems and the evaluation of significant environmental aspects.

The five Swiss case-studies are very different from each other from two points of view: the type and size of the companies and the stage of institutionalisation within the company. These five cases cover almost the whole range of LCA application in Switzerland, from small to big companies, from early to later stages of institutionalisation and from very simplified to complex methods used.

Four of the five cases are big industries which is not representative of the average of size of Swiss companies carrying out LCAs. Many smaller or medium enterprises have introduced this tool as we saw in the survey phase of this project. A big proportion of companies use a simplified tool, introduced by an association of "green" companies. The selected case studies do not represent this important part of LCA application in Switzerland, because none of these companies was willing to take part in this study.

5.2.1.2 Kraft Jacobs Suchard (KJS)

The company

Kraft Jacobs Suchard (KJS) is a leading European producer and marketer of a wide range of food products including coffee, cream cheese and confectionery (chocolate). KJS was created in 1993 by the combination of Kraft General Foods Europe and Jacobs Suchard and is part of Kraft Foods, the second largest food company of the world. The coffee-LCA underlying this case-study is the work of KJS Germany. In 1997, KJS Germany made a turnover of 3 500 million ECU and employed about 4 300 people.

As a processor of agricultural products, KJS itself is affected by environmental degradation. The company therefore has a vital interest in carrying out business in an environmentally sustainable way. Since 1992, environmental management systems have been developed, aimed at continually improving environmental performance. Of the KJS sites in Germany, a first site – the roasting factory in Berlin – has been certified according to ISO 14000.

The case-study

An LCA on was initiated at the beginning of the 1990s by an internal project group developing an environmental management concept for Jacobs Suchard. This concept postulated that a producer shares the responsibility for the entire life-cycle of its products and demanded co-operation with government and society (NGOs). The coffee-LCA was designed and worked out together with two partners: the Umweltsenator of Bremen and the Fraunhofer institutes at Bremen and Munich. Furthermore, a number of NGOs as well as a large retailer were invited to follow the study, a fact that promoted the study's credibility.

The coffee-LCA answered to the following questions:

- What is the minimal and maximal environmental impact caused by the production and use of coffee (worst- and best-case scenarios)?
- What is the effect of selected measures on this impact?
- How much do the various stages of the life-cycle of coffee (incl. packaging) contribute to the total impact?

The results of the study suggested that programmes aimed at reducing environmental impact should, as a priority, address the agricultural stage of the coffee production. Actions that could be undertaken by KJS individually respective autonomously proved relatively ineffective. As a consequence, KJS initiated for a co-operation project aimed at the development of best practice in the fields of cultivation and processing of coffee beans. Besides that, the coffee LCA triggered some spin-off measures (e.g. improved transport packaging in Sweden).

Present and future use of LCA

LCAs were introduced in KJS at the beginning of the 1990s. The first studies dealt with packaging only. The coffee-LCA was the first study that focused on the content of the packaging (a study on cream cheese followed afterwards). In general, LCAs were used to identify key issues of product design, e.g. areas for improvements of packaging respective to derive design criteria. The comprehensive LCA-studies concerning coffee and cream cheese were intended also to gather information with respect to the product's whole life cycle. KJS intends to carry out a further comprehensive LCA-study concerning confectionery, the third main product group (chocolate products). With regard to the three main product groups (coffee, cream cheese, chocolate), a rather systematic use of LCA can be observed. However, in general, LCAs are only occasionally used when the need arises. KJS seems to have gained a lot of experience with respect to the integration of environmental issues into day-to-day business management. However, the future role of LCA within KJS will most likely be of less importance (less means available, less comprehensive scope). The planned chocolate-LCA will almost certainly be the last comprehensive study.

5.2.1.3 Landis & Gyr Utilities (LGUT)

The company

Electrowatt is a group with several companies dealing with energy supply, energy and safety management and electronic goods. Landis & Gyr Utilities (LGUT) is a company in the Electrowatt group selling equipment for energy management (counters, valves etc.) and also conducting energy contracting. The annual turnover of LGUT is 625 million ECU (1996); about 5 300 people are employed.

LGUT's activities have to do with saving energy. Because these companies care about the environment with their activity, they have a certain obligation to optimise their product design and production processes in terms of environmental questions. For this purpose LGUT developed a tool, usable for designers to evaluate very quickly and easily their products.

Some companies in the Electrowatt group have already built up certified environmental management systems (ISO 14001 or BS 7750). An eco-efficiency report has been published in 1995 and 1996 covering several environmental aspects such as energy consumption, use of hazardous substances, air emissions or waste. In general these reports have shown decreasing environmental impacts over the whole group.

An important initiator for the introduction of environmental questions into business processes was the former head of the concern. He planned to establish environmental management systems in the companies; environmental engineers chose then LCA-methods for its implementation.

The case-study

Very few LCAs have been carried out. The one analysed was an LCA about a counter for electricity. This product was chosen because of its simple material content and the data available for the inventory stage.

The result showed, that the energy used during its life time is the most important impact. One reason for this result is the fact, that these counters have an extremely long life time. The production processes did not have to be changed because of this result. The application of the LCA showed in addition, that the tool is quite time consuming and not applicable for design processes. For this reason this result has been introduced into the tool for the product developing process (Product-Eco-Performance [PEP]).

Present and future use of LCA

In the near future, no further LCAs are planned to be carried out. One main reason is the insufficient data base for electronic goods. Another more important reason is the fact that LGUT will use its resources in developing a tool for product-designers which should be very easy to apply. This tool is called PEP (Product-Eco-Performance) which is a sort of checklist with environmental criteria such as: efficiency of material consumption, emission of hazardous substances, consumption of packaging, energy use, ability of recycling and management of product life cycle. These criteria are then rated and added together to obtain an overall score for each product, which describes the environmental performance of the product. The product designer is obliged to control his design process with this tool. He has to reach with his product a minimum of points and therefore has a clearly described goal for ecological product design. The fact that several criteria in the PEP are based on information of LCAs carried out in Electrowatt or elsewhere show the importance of this tool, even if it is not used continuously.

5.2.1.4 Ernst Schweizer AG

The company

Ernst Schweizer AG is a manufacturer of metalwork, it was founded in 1920 as a craftsmen company. Nowadays, it is a stock corporation, owned by the founder family, with an annual turnover of 40 million ECU and with 400 employees. The head of the Research and Development Department is also responsible for environmental aspects. The company produces and sells metal windows, facades, doors and mail boxes. In addition it develops solar energy systems. A lot of products of Schweizer are directly used as energy saving products, such as windows and doors with high insulation values or solar energy systems, which is an environmental engagement in itself. The company demonstrates its further environmental commitment also as a member of the Swiss Association of Environmentally Responsible Management (ÖBU). In addition the company has had an environmental management system, certified according to ISO 14001, since April, 1996. The production process itself causes very few environmental problems, since the painting facility has been renewed and improved with respect to air emissions of Volatile Organic Compounds.

The case-study

Two LCA-studies have been carried out at Ernst Schweizer AG, one about mail boxes, one about energy saving aluminium-glass folding-doors. The incentive for the LCA for mail boxes was a recommendation of environmentally oriented customers in public institutions who wanted to reduce plastic in general in building materials. Due to the fact that the mail box of Schweizer contains a rather high proportion of plastic, the sale of this product could be difficult in the future. Schweizer wanted to know whether products containing plastic really cause more environmental problems than others. For this purpose, two alternatives without any plastic were designed by Schweizer and were compared in an LCA study. These two alternatives contained aluminium and steel instead of plastic. The LCA clearly showed that from an environmental perspective the original product was the best alternative. The other two products would cause more environmental pollution due to the high amount of energy used for the production of aluminium or the heavy weight of steel. Schweizer was not encouraged to change the product design due to this result.

The second LCA was done for energy saving aluminium-glass folding-doors. This product was compared with an older product with less energy saving properties but with lower amounts of aluminium. The question was whether the energy saving effect of the new product would be more important from an environmental perspective than the higher amount of aluminium used for its production. The result of this LCA was unclear and a decision could not be made on this basis. As an SME, Schweizer was not able to repeat redo and improve the LCA, which shows the limit of this method for this kind of company.

Present and future use of LCA

In the future LCA will be used in a similar way as in the past. It will be used for the evaluation of products in the design phase or for the comparison of existing products with alternatives. It will not be used primarily for advertising and communication to the public but as an internal base for decisions.

The tool is rather complex for an SME, which can only actually invest about two man-months a year in this way. Schweizer wishes a simplification of the method, better access to data bases and more reliable software tools to enable an easier preparation of an LCA.

5.2.2 Germany

5.2.2.1 National situation

Germany is a country with a long LCA-tradition. The first LCA-studies were carried out in the 1970s (Oberbacher et al. 1974). The applications increased at the end of the eighties (see chapter 3).

The five German case-studies underlying this book cover companies of different sizes and branches: AEG Hausgeräte GmbH, Bosch and Siemens Hausgeräte (BSH) GmbH, Henkel KGaA, Weleda AG and YTONG AG. All firms are part of internationally operating groups of companies. Whereas BSH, Henkel, Weleda, and YTONG belong to the leading companies, AEG is a subsidiary of the Swedish Electrolux Group. Although the companies are all rather big, the number of employees varies considerably, from 500 (Weleda) to 47 000 (Henkel Group)[4].

The case-studies cover three different sectors: Two firms (Henkel and Weleda) belong to the chemical industry, another two (AEG and BSHG) are manufacturers of electrical appliances and one company (YTONG) produces building materials. As a producer of "classical" chemical products, Henkel is the only company that comes from an industry that was severely criticised as ecologically harmful.

It is known that especially the chemical and automobile sectors are the most active ones regarding LCA. But also the electrical branch is involved in several collaborative LCA-studies. Therefore, we think that the selected German case-studies are appropriate.

All companies sell their goods primarily to consumers (via retailers), e.g. the companies hold positions towards the end of the product life-cycle. The complexity of the single products respectively the range of products the companies supply differs significantly: Whereas all Ytong-products are made from the same basic material, the electrical household appliances produced by AEG and BSH consist of many different components. Of the LCA-studies we looked at, the study prepared by Ytong was the only one that analysed complete products; all other LCAs only dealt with product components. None of the firms belongs to branches which are known for frequent or radical product changes.

In most cases, a mix of internal and external factors has driven the companies to carry out the LCA-studies, external factors being the more important ones. The market, that is competitors and/or consumers, has been the decisive impetus in four of five cases (AEG, Henkel, Weleda, YTONG). Public measures like environmental legislation never triggered the carrying-out of the LCA-studies directly. However, external pressure was mostly accompanied by an internal interest to identify and reduce or prevent product related environmental impacts. Four companies actually carried out the LCAs, when a concrete product decision like the

[4] Numbers of people employed in 1996.

switch to another material or packaging had to be made (AEG, BSH, Henkel, Weleda).

Only one company (Henkel) regularly carries out LCA-studies; LCA is integrated there in the business processes and often used. Two firms (AEG, BSH) conduct LCAs from time to time, whenever somebody sees a need for comprehensive product related data. In the other two cases (Weleda, Ytong) the LCAs underlying our case-studies were and still are the only LCA-studies commissioned by those companies.

There are some tendencies on the future applications of LCA within German companies:

- Economic restrictions demand easy methods and "quick and dirty" LCA-studies, i.e. a clear tendency towards simplified streamlined LCA-studies;
- downstream companies tend to start with "easy" LCAs of components, materials and/or compounds and enlarge the complexity of their studies over time;
- the increasing importance of business-internal applications of LCA;
- the importance of collaborative studies especially among companies within the same sectors.

5.2.2.2 Bosch and Siemens Hausgeräte GmbH (BSH)[5]

The company

Bosch and Siemens Hausgeräte GmbH (BSH) was founded in 1967 when the Robert Bosch GmbH and the Siemens AG merged their household appliances divisions. The company produces a wide range of electrical household appliances like freezers, washing machines, and cookers and have meanwhile developed into an internationally operating group. In 1997, the BSH Group made a turnover of ECU 4.9 billion and employed about 32 000 people. The German market for electrical household appliances is dominated by a few well-known brands: With 33.8%, Bosch and Siemens Hausgeräte holds by far the biggest share, followed by Electrolux with 16.8%.

Traditionally, the household appliances industry has been relatively little directly involved in ecological problems. Most environmental impact results from the appliances' consumption of water, energy and detergents as well as from the high quantities of waste that occur, when the machines have to be disposed of at the end of use. Accordingly, BSH strives to reduce the water and energy consumption of its appliances; on average, both of these decreased by more than half since 1970.

At the end of the 1980s, BSH built up an environmental management system. As one part of the system the working group "Environmentally sound product development" was established, which co-ordinates the product related environmental

5 Until April 1, 1998, the name was "Bosch-Siemens Hausgeräte GmbH".

activities of the BSH Group[6]. This working group consists of representatives of all product related company divisions and is led by the employee of the central environmental department responsible for products.

The case-study[7]

The case-study looked at by us is based on an LCA that dealt with soap containers of washing machines, i.e. it compared steel soap containers to soap containers made from polypropylene, in order to identify the superior alternative. The LCA of soap containers was the first real LCA-study the company commissioned.

So far, the soap container in BSH washing machines had been made out of steel. In 1994, in the course of the development of a new top-loader washing machine, the product development department of the laundry product area thought of replacing the steel container with a polypropylene (PP) variant. These thoughts were not least stimulated by the fact that some competitors were already using soap containers made from synthetic materials. This planned switch was motivated more by economic and technical than ecological reasons. But because this switch meant an important and barely reversible change in the basic technology, the laundry product area decided that it should not be put into practice without considering the environmental consequences at the same time.

Thus in 1994, BSH commissioned an external institute to carry out an LCA that compared the two kinds of soap containers. The LCA-study was used in a prospective way, i.e. to support a planned decision; the study considered the complete life-cycle and weighted[8] the impact categories to an overall result.

The LCA-study confirmed that the container made from PP was superior to the steel variant for economic and technical reasons. With respect to the ecological question the result was less clear, but suggested that the PP-container was ecologically advantageous as well.

Subsequent to the LCA, BSH began to replace the steel soap containers. In 1996, the first washing machines equipped with containers made from PP were placed on the market. The LCA was hardly used for the identification of further optimisation potential or for the information of company internal and external stakeholders.

---

6 When BSH built up its environmental management system at the end of the 1980s, seven subject related working groups were established: for environmental protection related to products, production, logistics, distribution, procurement, communication, and co-operation with industrial associations. In the meantime, five of these groups have been replaced by environmental officers (co-ordinators), and only the working groups dealing with products and production have "survived".

7 The case-study is based on the short-version of the LCA (Saur 1995).

8 The weighting method is based on the Dutch proposal of the VCI-group (VCI 1994) which suggests to weight the environmental impacts according to their special extension (local, regional, global). The weightings actually used in the selected LCA-study were chosen by the external consultant.

By mentioning the LCA in the BSH environmental report, its existence was made known to a wider public. But the report on the LCA was restricted to a few lines, providing the basic facts about background, motives, and results of the LCA-study. The LCA was not applied in direct product marketing, mainly because it did not analyse a complete product, but only one part of the washing machine. Moreover the information provided by LCAs are generally regarded as too comprehensive and too complex to be used for public relations' activities or advertising.

Present and future use of LCA

The responsibility for the development of new or enhanced products lies with the single product divisions; each division has its own product development department. In order to generally integrate environmental aspects into the product innovation process ecologically oriented guidelines have been formulated. Among these guidelines, the obligatory tool called "Product-Environment-Examination" secures an early integration of ecological aspects in the course of the development of new product generations. Since the "Product-Environment-Examination" is meant to cover all life-cycle stages of the products, actual and target values could also be identified with the help of LCA-studies.

But so far BSH has refrained from firmly integrating LCAs in this process because conducting LCA-studies of complex products like electrical household appliances is regarded as almost impossible. Thus the "Product-Environment-Examination" covers the products' entire life-cycle, but regards only selected energy and material flows. Moreover, BSH does not systematically integrate LCAs in the product development process, because this would require that LCAs accompanied the whole process and were continuously adjusted to its advances. BSH regards the instrument as too comprehensive to be used in this rather dynamic way. LCAs can only depict momentary situations, i.e. they are valid only for special scenarios set in particular studies.

Because BSH has experienced the conducting of LCAs as very demanding, the instrument has not been applied regularly, but only to support important product decisions. According to the environmental department of the laundry product area, "important" means that the appliance's technology, design or operation change remarkably. Moreover, the company regards LCA as not flexible enough to systematically accompany dynamic product development processes. However, the laundry product area is still the only product area showing interest in carrying out LCA-studies. The only study the company performed subsequent to the LCA of soap containers was also conducted there.

No urgent need for further comprehensive life-cycle studies has been seen, because it is meanwhile well-known that most environmental impacts (about 80-90%) related to electrical household appliances occur in the stage of use. Thus, the focal point of BSH's research activities are already defined. Moreover the company claims that comprehensive data on the eco-performance of its products already exist. BSH uses LCAs to identify ecological weak-spots and to support important product decisions. So far, BSH has not carried out LCAs of complete appliances, but analysed single product components or packaging systems.

The company does not plan to carry out LCA-studies regularly, and there is no intention to build up a comprehensive internal LCA-know-how. The conducting of any future LCAs will again be laid in the hands of external experts.

5.2.2.3 Henkel

The company

Henkel is one of the biggest producers of chemical products world-wide. Its products ranges from adhesives to washing and cleansing agents. Henkel has gradually developed into a globally operating group of companies. With 246 associated companies the Henkel Group is present in more than 60 countries. The leading company is the Henkel KGaA, public limited partnership, in Düsseldorf. In 1996, the group made a turnover of ECU 8.2 billion and employed 47 000 people.

The company claims a leading position in the field of environmental protection. Accordingly, it constantly works on an ecological optimisation of products and substances, in the sense of a comprehensive "product stewardship". In this context Henkel has been using LCA from the beginning of the 1990s.

The case-study[9]

The need and the willingness to carry out an LCA arose firstly in the division of washing and cleansing agents. In the late 1980s, it was decided to substitute the customary petrochemical surfactant (LAS) for a surfactant with an oleochemical basis (FAS) which was supposed to be ecologically advantageous. But shortly before the market introduction, a competitor published an LCA in which LAS came off better than FAS. Subsequently Henkel planned to commission an LCA itself, in order to be on the safe side. This is the LCA-study we have already examined.

The responsibility for conducting the LCA-study was placed with the department WEQ-Quality and Environment of the washing and cleansing agent's division. The study was carried out in close co-operation with the division of chemical products, because important detergent ingredients are produced there. In order to co-ordinate the internal co-operation, a working group of ten to twelve employees was installed which included representatives of marketing, R&D, as well as natural scientists.

The study itself compared LAS and FAS. It was the first LCA-study carried out by Henkel. Accordingly, it was related to great effort in data collection and meth-

---

[9] The case-study is primarily based on two publications of Henkel's employees (Klüppel 1993, Klüppel et al. 1995).

odology learning. Although Henkel made enormous efforts, the company conducted merely the most important LCA-steps and regarded only the relevant life-cycle stages.

The (cradle to gate) LCA came to the conclusion that the oleochemical surfactant FAS had to be preferred for ecological reasons, although it was not superior to its petrochemical alternative LAS with respect to all impact fields. Besides the comparison between the two surfactants the LCA identified ecological weak-spots of FAS, but triggered hardly any optimisation measures. The results of the LCA were used more to document and communicate the existing advantages of FAS.

Although efforts have been undertaken, a complete substitution of LAS by FAS has not yet been realised. After the detergent version of 1993 contained exclusively FAS, in the version of 1995 a mixture of FAS and LAS was again used. Nor could an entire replacement of LAS be reached in the course of the innovation of "Persil Megaperls". This was firstly due to technical problems. They have been solved in the meantime, but the cost aspect still remains. Due to a massive cut in the price of LAS, FAS has simply become too expensive. Due to the very price-sensitive demand in the detergent sector, consumers could not be induced to pay more for ecologically optimised products.

Within the company the results of the LCA have been transmitted especially to decision-makers responsible for product management. The LCA influenced the product development in so far as it confirmed the company in its efforts to switch from LAS to FAS. Since the results were introduced in the public discussion about pros and cons of FAS and LAS, the LCA was also used in the fields of external information and communication. Information on the LCA were distributed through lectures and articles in specialist journals as well as through environmental reports and brochures of the company.

Present and future use of LCA

While the first LCA resulted from external pressure and primarily served external information and communication purposes, nowadays the internal use of LCAs is to the fore. They are firmly integrated in the continuous ecological optimisation process of products and substances. Additionally LCAs serve to provide information for consumers and stake holders in the context of public relation activities. Because LCA-studies often reveal complex and indefinite results they are not used as instruments for direct product marketing.

Within the product (and substance) innovation processes of Henkel in the area of washing/cleansing agents some typical steps have to be carried out[10]. When the product idea has turned into a real product, this is comprehensively tested on its ecological, biological, dermatological, and toxicological properties. The decision about the biological admission of products is taken in accordance with minimum criteria that are set by law, voluntary agreements or Henkel internal directives. To

[10] See Figure 7.2 in chapter 7 for the formal procedure.

determine whether substances and products fulfil the minimum internal criteria an analysis by means of negative lists, checklists and risk assessments takes place. Additionally LCAs are used for analysing substances and single products or to compare different product alternatives. The relation between LCA and other instruments is more supplementary than competitive. Being a relatively complex instrument, LCAs are only used after the tests with the help of other instruments have reached positive results. In the division of washing and cleansing agents the division management has given instruction to systematically carry-out LCAs for every "important" product innovation.

Because the tests are conducted in the field of product development, the developers are the first to realise a need for an LCA. Staff members in charge of conducting LCAs, e.g. the WEQ-department in the washing and cleansing agents division, are informed or are sufficiently integrated in the product development to become active by themselves.

The complex LCA-results are summarised and made available to decision-makers in management and product development. They are inter alia presented in the form of the so-called "collection of substance data". This data sheet encompasses the most important substance related information in a very condensed form. Selected life-cycle data (cradle to gate-LCA), basic data of the ecological and toxicological analysis and other relevant data are depicted.

In the beginning of its LCA-activities, Henkel closely co-operated with external experts and consultants, because the company internal LCA know-how was not sufficient. Meanwhile Henkel is well able to carry-out LCA-studies independently. Henkel built up its own data bank from the start of the LCA-engagement; this turned out to be very helpful. Due to the possibility of using data modules provided by the data bank, the time needed for conducting an LCA has been reduced considerably. Often an LCA can be commissioned within a few days. The data bank is continually updated and enlarged. Currently it contains data modules of 250 different processes and of over 400 substances.

Personnel capacities for conducting LCA-studies were not only built up in the washing and cleansing agents division. The sectors of chemical products, metal chemistry, and packaging are also engaged in LCA. In the other operative divisions a more intensive use of LCA is prevented by data problems.

All in all, LCAs constitute a useful completion of Henkel's environmental management instruments. Therefore the company has stated that LCA will, in the context of other instruments, play an even more important role in the future.

5.2.2.4 Weleda

The company

The history of the Weleda AG, a Swiss public limited company, began in 1921 in Arlesheim (Switzerland). Its philosophy is rooted in the anthroposophical world view of Rudolf Steiner. Accordingly, it always aimed at producing in harmony with humans and nature. Steiner was decisively involved in the foundation of the company.

The company produces naturally based pharmaceuticals and products for personal hygiene. On the markets for these products Weleda holds a strong position. In 1996, the 476 employees of the German Weleda location in Schwäbisch Gmünd made a turnover of ECU 60 million and thus contributed by far the biggest share to the whole international company's sales.

Since Weleda uses naturally based ingredients, its products can be regarded as ecologically harmless. Environmental impact are arises more from production or packaging. For example, Weleda followed the general trend towards smaller packaging which resulted in rising packaging intensity. Most of Weleda's pharmaceutical and cosmetic products are filled in traditional aluminium tubes and glass bottles, only in recent years did ithe Weleda works in France and Switzerland begin to switch to synthetic materials. The "LCA of packaging systems" that was commissioned in 1995 revealed environmental advantages for the synthetic materials.

The case-study[11]

The German public, consumers and state show a relatively critical attitude towards the packaging problem. For Weleda Schwäbisch Gmünd the reason for commissioning the "LCA of packaging systems" was to analyse the ecological performance of the glass bottles and aluminium tubes traditionally used and to compare them to synthetic packaging alternatives made from polyethylene (PE) and PET. With the help of the LCA the environmental impact related to the different packaging systems was to be registered, evaluated, and compared, in order to identify the ecologically most desirable alternative.

In 1995, an external institute carried out the LCA. The LCA delivered a comparison between already existing packaging. In this respect, it is a descriptive study. But the LCA was not only used for a documentation of the environmental impact related to the different packaging systems. It also served to identify the ecologically most desirable alternative that should be applied in future. In this respect the study had a prospective function for the company. The whole life-cycle stages of the two packaging systems were regarded and all LCA-steps were carried out.

The LCA revealed significant ecological advantages for the PE-tube over the aluminium tube. With respect to the bottles the result turned out less clearly. Although the PET-bottle came off as the ecologically most desirable variant, the brown glass bottle was recommended by the external institute that carried out the LCA, due to the better disposal situation.

[11] This case-study is based on the LCA-study (Ankele/Steinfeldt 1995).

According to the LCA-results, Weleda wants to replace the aluminium tubes as well as the glass bottles by the synthetic variants. But for reasons of product quality and preservation a switch was not possible without adapting the products' formulas. This has not been realised so far, although the product development department has made intensive efforts. So far, the LCA-study has served to improve the traditional packaging systems, for example, by reducing the weight of glass bottles or by not using cardboard boxes and instruction leaflets.

In the environmental report of 1996, a goal was set to develop a new packaging concept for tooth pastes with a switch from aluminium tubes to PE-tubes and with no further use of cardboard boxes.

The LCA was commissioned to enable the company to give scientifically based answers to critical consumer questions on packaging. In this respect the LCA also served for external information and communication purposes. This information was not provided in the form of advertisements, but in the course of public relation activities. LCA-information was made available in the Weleda-news or in response to concrete requests.

Present and future use of LCA

So far, the "LCA of packaging systems" is the only "real" LCA that has been conducted by Weleda. Additionally some studies have been carried out which show LCA-characteristics, but cannot be regarded as complete LCA-studies.

All LCA-studies or LCA-similar studies conducted by the company have not analysed Weleda products, but input products (packaging materials) obtained from suppliers. Accordingly LCAs are more used in the context of procurement than in product development. The "LCA of packaging systems" has influenced product development only indirectly, in so far as a switch from aluminium to PE-tubes requires modifications of the products' formulas. Currently LCAs are not systematically integrated in decision-making processes, but are only occasionally used.

In general Weleda regards LCAs as a useful instrument to include ecological aspects in the development of products. But with respect to its own products the company currently sees no need for action, because Weleda products are based on natural substances and thus are mostly environmentally harmless. In the product development of Weleda, other information providing instruments like technical check lists, evaluations of suppliers, and market surveys play a more important role.

In the long run the company plans to carry out LCA-studies of its own, particularly new, products as well. In this way it aims at a more systematic integration of ecological aspects in the product development process. Although it has always tried to consider environmental impacts, so far this has been done in a rather subjective way. Nevertheless, at present not all parties involved see a necessity to regularly commission LCA-studies.

Weleda regards its experience with the "LCA of packaging systems" as predominantly positive. The provision of a systematic overview on all environmental impacts related to products is regarded as the main strength of LCA. The most important weaknesses of the instrument are seen in the comprehensive data-collection and the manipulability of the results arising from an incomplete standardisation.

5.2.3 Italy

5.2.3.1 National situation

In Italy, the use of LCA has begun only recently.

The five Italian case-studies regarded differ very much from each other from two points of view: the type and size of the company and the stage of evolution of LCA within the company. In fact, they range from the case of FIAT, the largest Italian company, to Cartiera Favini, which is a small-medium enterprise (SME).

Only FIAT is strategically and organisationally well placed and fully committed to the use of LCA as a routine supporting management tool. Most other LCAs carried out in Italy until now have been pilot-projects. However, several companies are planning to make a much wider use of LCA in the future. It is evident (in three out of five case-studies) that the influence of a more advanced corporation, located outside of Italy, has played a major role for starting and advancing forward the use of LCA. The case of Cartiera Favini is a "special", story, because it represents a "fully Italian SME case", which differs completely from the other cases in all aspects.

Another trend, which emerged in our survey, is the tendency to use LCA in simplified forms. This might imply the cutting out of the Impact Assessment phase, and a more systematic use of screening LCA. In one case (Italtel), this meant also developing LCA-alternative tools in the form of semi-quantitative environmental indicators.

As a consequence, the application patterns of LCA in Italian companies are expected to develop along two distinct lines:

- A large use of simplified LCA (eventually combined with the use of other environmental management tools), in order to get a flexible support tool for Research and Development and Product Design activities.
- Fewer applications of more complete LCAs, for final products and their Processes,

In general, today (and in the near future), the application of LCA is limited as an internal tool for R&D and product design, for two main reasons:

1. The Italian market is not yet considered to be sufficiently sensitive to environmental concerns.

2. Companies generally think at least at present that results of LCA are too complex, and often disputable, to be communicated to the public.

There is a common feeling that LCA might be used also for marketing purposes in the future. However, this requires the achievement of a certain degree of standardisation of the methodology, quality of data and ways of reporting.

In none of the Italian case-studies LCA has been directly encouraged or motivated by legislation or public policy. In fact, there is no framework legislation on environmental issues in Italy. Very often single regulations and roles of public actors are confused and overlapping. Only the recent (1997) waste act is explicitly related to the final phase of the product life-cycle and recommends the use of LCA as an analysis tool. However to date, companies are concerned only with the packaging of their products.

On the other hand, companies think that there might be an increasing correlation between the European Directives and the use of LCA in the future (FIAT is an example of this anticipation of European regulations). However, market pressures are perceived to be generally more important, now and in the future, than legislative pressures.

There is a common feeling about the lack of a publicly available, shared and reliable data-base on infrastructure sectors (energy, transports, waste) and materials. The National Environmental Agency (ANPA) is currently responding to this need, as it has commissioned an important LCA study, which should precisely serve as reference National data-base in the future.

Finally, it is worth mentioning that the Italian Association for LCA was officially founded in January 1999. The association is meant to be a forum for discussion on LCA issues and experiences. Beyond companies, the Association will include representatives of public authorities, research institutes, universities, consultants, NGO's and consumer organisations. To date, among our case-study companies, FIAT, ABB, and ITALTEL have agreed to participate.

5.2.3.2 Fiat Auto

The company

FIAT Auto is the main company of the FIAT Group operating in its "core-business", that is the production of transportation vehicles[12]. The FIAT Group operates in 60 countries with 874 companies. In 1996, the group had a turnover of around 40 billion ECU[13], of which 80 % were related to the core business. 37% of its production was sold in Italy, 40% in the rest of Europe, and the remaining 23%

---

[12] A few companies of the FIAT group other than FIAT Auto are involved in LCA activities as well. However, until now, LCA has only been applied to cars and/or car components.

[13] 1 ECU = 1 947 Italian Lira.

outside of Europe. In the same year, FIAT had 11% market share for cars and 20% for industrial vehicles in Western Europe. FIAT Auto has the largest share of the automotive market in Italy. It has also a leading position at the European level.

In 1991, FIAT Auto set up the "Direzione Ambiente e Politiche Industriali" [DAPI][14], in order to tackle the complex issue of assessing and reducing the environmental problems of cars. The department was also established to respond to the emerging and increasing public attention connected with the 1992 Rio Conference on Environment. This department is responsible for both environmental and industrial policies. DAPI is well integrated in the decision-making structure of the company. It occupies the same level as other important departments, such as the technical, purchasing, and production departments. Its head refers directly to the chairman and chief executive officer.

The environmental policy at FIAT Auto is defined in a long-term strategic framework programme called "Project 21", which takes its name from the Agenda 21, defined at the Rio Conference, and which was at DAPI. As a core tool for the wider concept of Life-Cycle Management, LCA is explicitly included in the "Project 21".

FIAT certainly is the company with the widest experience of use and application of LCA in Italy. Since November 1997, the firm has completed 16 LCA studies. Two more are being carried out at present. Many others are planned for the near-mid future.

The case-study

Among the many possible LCA studies, we selected together with the interviewees the one related to the comparison of two equivalent engine blocks, one made from cast iron and the other from aluminium. This has been one of the most significant studies carried out at FIAT for several reasons. Firstly, it is related to a single, but very central car component, that is the engine. Secondly and most interestingly for our case-study, it led to rather surprising results. In their turn, these led to concrete decisions which changed, at least partially, the company's former strategy with respect to this particular matter.

In fact, just like many other car producers, FIAT has embraced the strategy of reducing the weight of its cars as much as possible with the goal of reducing fuel consumption. A major car component "target" for this strategy is the engine block, since conventional cast iron engine blocks can be substituted by lighter ones made entirely from aluminium.

However, previous LCAs had introduced some doubts about the actual consequences of generic environmental strategies. The LCA comparing two different materials for engine blocks was carried out with the explicit objective of verifying whether the generic strategy of reducing weight was a correct environmental solution from a life-cycle perspective or not.

---

[14] DAPI means "Department of environment and industrial policies".

The results were surprising, at least to management people not used to considering the whole life-cycle of materials. In fact, before this study, it was generally thought that any car weight reduction must be environmentally friendly. The company strategies with respect to car consumption reduction had been tuned according to this "generic principle". Instead, the study demonstrated that the use of alumimium can be environmentally preferable, but only under certain conditions. The study concluded that, under the current state (1996) of the F.A.RE. system[15] and of the Italian market of secondary aluminium alloys, the use of aluminium on a large-scale must be considered with great care.

This had significant implications on the strategy of the company. In fact, after the study, the decision was taken to slow down the substitution of cast iron engine blocks with aluminium blocks.

The study did not reject the aluminium solution in general, but it suggested to review conclusions in the case of the evolution of external parameters, such as the actual number of recycled cars and the availability of suitable secondary aluminium alloys on the Italian market. It was also from this perspective that aluminium engine blocks were actually applied in a new car model (Alfa Romeo 156). Today, this is still a niche market application.

It is worth highlighting that these decisions have been taken mainly for economic reasons and not only for ecological ones. The full substitution of cast iron engine blocks on all existing models would have implied major production changes and consequent investment. The environmental analysis further supported this decision. According to the interviewees, a manufacturing extra-cost for the environmental improvement of the car and its components is allowed in principle. However, in messy situations, economic arguments always prevail.

The selected LCA study was a prospective one, carried out as a decision-making tool, and its application was on a large scale. It was a full LCA study; all product life-cycle stages and LCA phases were been included. Several impact assessment methods and software were used for the comparison of results (mainly for learning purposes).

Both the staff of FIAT Auto (Department For Environment and Industrial Policies) and of FIAT Group (Research Centre) were involved in the study. There was also minor external support. However, now all LCAs are carried out internally.

Present and future use of LCA

The interest of the company in LCA is very strong. There is a clear commitment to further expanding its use and to integrating it as a routine tool into decision-making processes. LCA will be a major supporting tool within the framework of the forthcoming (1998) "Integrated Development Plan of (car) Components". This plan, which crosses through several departments (apart from DAPI, it will involve

---

[15] F.A.RE. means "FiatAuto Recycling", the system implemented by Fiat to recycle dismissed car materials.

the technical, purchasing and production departments), is going to define the policy of FIAT with respect to material choices and design of car components. It will include a new specification introduced for designers, that is the environmental impact of materials and of project alternatives. It is from this respect that LCA is going to be increasingly used in a systematic way. In general, the use of full LCAs is expected for materials and car components, whereas streamlined LCAs are and will be used for more complex issues (e.g. electric car, bio-fuels)

The results of LCA are currently applied only internally. In particular LCA is not perceived as a marketing tool, because of its complex and sometimes controversial results. As for other environmental external communication and initiatives, FIAT focuses on target audiences, such as schools, Universities, and experts rather than on the general public.

No LCA study has been directly triggered by legislation pressures. However, LCA is considered to be a very useful tool for anticipating legislation trends, particularly at the European level (i.e. it *is* also used to anticipate the forthcoming EU-directive on recycling of car components and materials).

There is clear evidence that LCA activities at FIAT are very well integrated into the process of product innovation. There are several good reasons for this:

- The Environmental Department DAPI is responsible at the same time for FIAT Industrial and Environmental Policies and is well integrated in the over-all organisation scheme of the company.
- Apart from LCA and environmental matters, there is a high level of integration between the different departments because of a "vertical", product-oriented organisation.
- LCA activities are coordinated by the Central Laboratories of DAPI, which have the responsibility to communicate and discuss the results with the other departments. This is a key-point, since the Central Laboratories have always been responsible for providing data on materials to the other departments and to the management. This has the major implication that the other departments (Technical, Purchasing, Production, etc.) receive all information, including the environmental data and LCA results, from the very same officers from whom they have been used to receiving basic data on materials. This is a very different situation from receiving LCA data from an environmental dept. far away from the traditional decision-making process.
- LCA activities will further increase and be integrated into the "Integrated Development Plan of car Components".
- LCA "culture" has now spread out into all Design & Development departments.

The following main implications / requirements for the wider use of LCA have been identified at DAPI:

- The use of full LCA for the analysis of single components, and the use of simplified LCA in the case of the whole car. The need for agreed simplification rules.

- The need for a public data-bank on energy systems, transportation, waste systems and materials, in order to be able to compare results and data coming from different sources.
- An increased commitment towards a careful supply-chain management. The collection and quality of data from suppliers is still a problem. The issue is how to involve and motivate them.

The real challenge that FIAT is going to face is the integration of LCA-thinking at all company levels and functions. In fact, FIAT has developed significant know-how with respect to "technical" aspects of LCA and the methodology is to be integrated more and more into design and development choices. This process involves all technical departments, is supported by the top management and will even accelerate through the implementation of the "Integrated Development Plan of car Components". On the other hand, the marketing, commercial departments and the network of dealers are still excluded from this process.

5.2.3.3 ABB ITALY

The company

ABB is a global Electrical Engineering Group, with an annual turnover of around 27.7 billion ECU and over 210 000 employees distributed in three world regions: Europe, the Americas and Asia Pacific. ABB's activities are organised in four business segments: Power Generation, Power Transmission and Distribution, Industrial and Building Systems, Financial Services.

In Italy, ABB has acquired several Italian companies: the biggest is SACE, to which our LCA case study refers. SACE has three industrial sites and represents about one sixth of the whole Italian ABB group with over 2 200 employees and an annual turnover in 1995 of around 1.8 billion ECU). Its main products are low and medium voltage circuit breakers for industrial and public sector use.

As a group, ABB claims a leading position in the field of environmental protection and describes itself as a pioneer. The organisational structure confirms this high environmental concern, as the Environmental Affairs function is placed directly at the Senior Vice Presidential Level and controlls a network of 43 national Corporate Environmental Controllers (CECs). CECs are charged with sending out Corporate environmental directives to all the countries ABB operates in. Environmental Policies at the corporate level have been established following the Rio Summit in 1992, since when the Corporation has been tackling the affiliated companies to subscribe to them and take action accordingly.

At the corporate level the environmental policy has recently evolved from focusing mainly on production processes to products themselves. The need to take account of environmental aspects related to products and to implement an LCA came directly from the very top management of the group. It was motivated by a number of reasons such as:

- the evolution of market environmental concerns and environmental policies from process to product focus;
- changes in ABB's strategies shifting towards a stronger "outsourcing" of production processes;
- the type of products: in fact, most environmental impacts of ABB's products are to a very large extent due to product operations (use and end of life) rather than to production itself.

ABB SACE was the first Italian company to take this concern seriously by starting an LCA, supported and encouraged by the Italian based ABB Ricerca, one of the Corporate is nine research centres throughout the world.

Amongst other "traditional" research and experimentation activities, in fact, ABB Ricerca started supporting ABB Italian companies in LCA, following Corporate directives on both the methodological and organisational aspects.

The case-study

The analysed LCA study was the first LCA experience in ABB Italy and also a pilot project. The study, proposed to ABB SACE by ABB Ricerca, compared the environmental impact of the new model of a low voltage circuit breaker with its old version, which has been on the market since 1988.

The LCA was mainly retrospective, aiming at confirming former decisions made during the project stages of the new circuit breaker, such as increased efficiency in resources use during production and the reduction of energy losses during use phase. Although it was conducted during the late stages of the product development, no changes were introduced in the final new product on the basis of the LCA results.

At that time (1996) none of the ABB Italian companies had any internal competence in LCA. At the corporate level however there was some experience: The Swedish Research Centre, had started testing LCA from the early 1990s. The Swedish Centre transmitted its LCA know-how to ABB Italy for the first LCA study.

The conducted study was a "full" LCA: all life-cycle stages and LCA steps were taken into account.

One person at ABB Ricerca was the co-ordinator and responsible for the whole LCA study. He also personally performed some elements of the analysis. Other people were involved at SACE, namely the Department of product development and design, directly responsible for the project stages of the low voltage circuit breakers under study.

ABB Italy's CEC was a crucial supporter of the LCA; its role consisted of encouraging ABB's Italian management to start LCA application following corporate directives.

The LCA confirmed substantially the choices made from an environmental point of view, ones that were quite clear before the LCA study anyway. In fact ABB designers had already taken into account environmental optimisation concerns while designing the new product, although no specific methodology or guideline had supported these choices.

Just one surprise emerged, due to largely the methodology chosen for the evaluation phase[16]. In fact, the new model rated worse than its older version in the manufacturing stage, in particular because of the indicator "resource depletion", due to the use of silver that was not used in the previous model. However, thanks to this additional component, the power losses in the use phase decreased significantly compared to the old circuit breaker, and the use phase is the crucial one for the life-cycle environmental impacts of circuit breakers. Therefore the new product rated globally better than the old one.

Moreover the LCA also identified a few weak spots in the new circuit breaker life cycle and potential optimisation opportunities that could lead to cost reductions.

Present and future use of LCA

The ABB case study gives an example of how a company can get started on LCA and how the whole organisation reacts to it in terms of new roles, new issues discussed and new points of views raised.

In fact, although the LCA examined was a rather retrospective / learning study, it had important implications and led to various changes within the company. On one hand, the study simply demonstrated and confirmed the superiority of the new product with respect to the old one also from an environmental impact point of view. No practical decision related to the product was taken. On the other hand, it led to several strategic decisions and to some changes in the organisation such as the creation of a specific LCA function in ABB SACE. Moreover, other Italian ABB companies have begun to deal with LCAs in the meanwhile. An inter-company working team for the development of LCA within Italian ABB companies has been established. It is coordinated by the responsible person for LCA in ABB Ricerca.

Although today the implications of LCA results on business decisions are still limited and product innovation still remains market driven, LCA is given a strategic value at ABB and is considered as the correct tool to assess the environmental impact of products.

ABB intends to expand the set of applications of LCA in the future and to apply it in a more prospective way. However, the company will most likely use full LCAs only for a restricted range of products, whereas simplified LCAs may be more easily integrated in the development processes for all new products.

[16] The "Environmental Priority System" (EPS) used in the Swedish Research Centre is very severe from the point of view of resources depletion (see Ryding/Steen 1991).

5.2.3.4 *Italtel*

The company

Italtel is a large industrial group supplying telecommunications networks throughout the world, both to private and public clients. Until a few years ago, Italtel was for the major part controlled by the Italian state owned company STET; after privatisation, Italtel merged in 1996 with Siemens Telecomunicazioni, an Italian company belonging to the Siemens Group, thus widening its activities and markets. At present Italtel has access with its full-line offer to more than 100 countries. This offer ranges from the supply and installation of a single piece of equipment to field assistance for the management of large communication networks.

In 1996, sales on the Italian and international markets reached two billion ECU, around 38% of which came from exports. Italtel's personnel at the beginning of 1997 accounted to 15 892, after a reduction of around 2 000 that took place after the merger.

Italtel's initiatives in environmental management area are quite recent, and received a strong push by the above-mentioned changes which occurred in the business context. Italtel adopted an environmental policy in 1994.

The question of the compatibility of telecommunication products with the environment started to emerge both because of the market push, such as through clients asking for product environmental information, and because of "institutional" encouragement, such as the evolution of the CE Label requirements for electronic and electromechanical products.

Environmentally related activities and initiatives at the corporate level have always been carried out within the Central Engineering Department. A separate function was created with the task to carry out research on new materials and technologies for telecommunication products. This department, called "Ecocompatibility and New Technologies", leads and co-ordinates both the LCA activities and the development of a specific internal assessment methodology of product environmental impact which was also analysed during our case study.

The case study

This case study refers to a specific methodology the company has developed as an alternative to LCA in order to assess the products' environmental impact. This methodology is called "Eco-compatibility Analysis", and the reason for focusing our case study on it is its integration in business decision making processes. In contrast, LCA, at the time the case study was carried out, was performed at a pilot project stage only with no specific impact on company decisions.

Here we summarise the shift Italtel made from LCA to the Eco-compatibility Analysis.

The Central Engineering Department, and in particular the function for "Ecocompatibility and New Technologies", came across LCA while assessing improvement environmental potentialities of materials and processes. Not having any internal skill at that time, an LCA was commissioned externally from an Italian research institute.

This first experience, carried out in 1995 on the basic telephone packaging, was intended to introduce a "learning by doing" process into the company for the possible use of the methodology internally. Therefore, the work was divided between the consultants and the internal personnel. In particular, Italtel's personnel gathered all internal data, whereas the consultants collected the external ones (related to transport, suppliers, energy use etc...) and carried out the other steps. Despite the positive approach to LCA, this first experience convinced the management of the high degree of complexity of this tool and of the uncertainty of specific results in the assessment phase.

The conclusions, at this stage, were that LCA did not seem appropriate for use within the company's product innovation processes. In fact, the internal data for the inventory construction corresponded to just 10% of the overall data needed for the completion of the whole life cycle. Besides, Italtel's management thought the assessment and evaluation steps were sometimes ambiguous, apart from being very complex.

The conclusions about the LCA drawn by the management were that the methodology did not satisfy the company's needs because:

LCA is too complicated to be manageable within a product's decision making processes;

A complete LCA involves a wider world than the one the designers can have influence over.

Italtel felt that smaller and more controllable boundaries should be selected to study ways of improving the environmental performance of its products. The company felt the need for a tool closer to the designer's culture, as well as complying with emerging market needs. For this reason the Eco-compatibility and New Technologies department started to define an internal and alternative tool.

The methodology that was elaborated, called "Eco-Compatibility methodology" aims at giving a measure of the environmental compatibility of a certain product. In brief, the "Eco-compatibility Analysis" consists of an environmentally oriented evaluation of the needed materials, the project solutions adopted and the fabrication and logistic processes applied to the product. The picture of a product's level of eco-compatibility, is built up thanks to several different indexes, and finally summarised in a general one.

The methodology is clearly designed for implementation in daily routines for product innovation: Thanks to the assumptions made for the construction of indexes, the methodology allows product designers to focus on those environmentally negative aspects of products upon which they can have an influence, whereas those negative aspects that are not under the direct influence of designers or of the company rate lower. This methodology considers just the product life cycle phases that can be directly controlled and managed by the company. It is a "gate to gate" assessment.

The "Eco-compatibility Analysis" was immediately applied as a tool for Design for Environment within the late development stages of a new product in Italtel's L'Aquila site, in central Italy, under the supervision of the Central Engineering department.

In L'Aquila, the activities are related to the research, testing and production of multiplex systems, cross-connections and access systems on optical fibre; mobile radio networks and defence telecommunications. In particular, the product subject to the Eco-compatibility methodology was a piece of electronic equipment that has the function to serve and link together different telecommunication centres. This equipment was developed by Italtel in L'Aquila and its release on the market was planned for the early months of 1998.

The Eco-compatibility Analysis took around three months and spotted a few environmentally weak points of the product that were partly reviewed before the definitive completion and industrialisation.

The methodology study and application involved four people belonging to the Engineering and Quality Departments and was well accepted, thanks to its user friendliness and its closeness to the Company's project and design culture.

Present and future use of LCA

As mentioned, Italtel made a strategic choice after testing a full application of LCA. In fact it decided that a simplified version of LCA would better respond to the company's needs, as far as product design decision making processes were concerned.

LCA has remained at a National Corporate level where new studies are still being carried out within the Central Engineering department; in particular two more LCAs have been done since 1995 and the company is internalising specific LCA skills more and more. Italtel's strategy is to continue assessing and testing LCA methodology while building an internal know how.

However, it is very likely that LCA will remain a learning tool mainly at R&D level for the environmental screening of products, used and discussed for a restricted number of functions at the Central Engineering or top management level. On the other hand, the "Eco-compatibility" methodology will increasingly be included in the daily activities of designers of the Group.

Although LCA is not directly used for implementation in the environmental product innovation process, its results are still taken into consideration for providing framework guidelines. The strategy is both to better define the criteria used in the Eco-compatibility Analysis, and to provide management with additional scientifically-based information on the environmental impact of its products. In particular, after the first trial test, the Eco-compatibility Analysis became an internal procedure, shifting from its previous guideline status.

Two main elements that would ease the future use of the Eco-compatibility Analysis emerged in this case-study: the construction of an internal data base for data collection and the creation of a function explicitly charged to supervise and co-ordinate all environmental related product activities within the company.

In our opinion, even if from a methodological point of view weak spots in the Eco-compatibility Analysis can be easily pointed out when compared with a full LCA, nevertheless, at the moment this methodology is strongly supported by the management. It is very likely that, in the near future, this will be the most used tool in Italtel's product environmental innovation processes. There is clear evidence that the company embraces the concept of Life Cycle Thinking, but it needs more simplified tools to translate this approach into implementation routines for product environmental innovation.

5.2.4 Sweden

5.2.4.1 National situation

The history of LCA in Sweden started in the 1970s when Tetra Pak as one of the LCA pioneers made a study of milk packaging. However the use of LCA was limited during the 70s and 80s. LCA activities dramatically increased in the early 90s. The government ordered a packaging study (Tillman et al. 1991); EPS was developed by Volvo in collaboration with IVL (Institutet för Luftvårdsforskning) (Tillman 1998), and the Product Ecology Project was started.

The Swedish industry has played an active and driving role in the Swedish LCA history. In short, their intention is to take the initiative and not let the government lead the environmental debate (Ryding 1998). Within the Product Ecology Project, methodological issues (allocation, system boundaries, LCA's role in decision-making) were pursued. The Product Ecology Project, which was coordinated by the Swedish Federation of Industries, involved 15 Swedish companies and research organisations; in addition, it contacted actors who have been important for the further development of LCA in Sweden. These contacts led to the establishment of a national centre of excellence in LCA and product-related environmental assessment, CPM[17] (Centre for environmental assessment of product and material systems) in 1996. However, this does not mean that Swedish industry has been working with LCA "in isolation". In contrast, there has been support from funding bodies. CPM is jointly funded by industry, The Swedish National Board for Industrial and Technical Development (NUTEK), and Chalmers University of Technology. Scandinavian management is characterised by, among other things, flat hierarchies, co-operation and informal channels (Jönsson 1996). This goes for the LCA activities in Swedish industry as well. The overall goals of

[17] See also Box A.

CPM are is to work for a long-term build up of knowledge that is relevant to the needs of industry in the field of product-related environmental assessment in order to prevent and decrease environmental impact associated with products. A major instrument in reaching this goal is the establishment of an LCA data-base. For this purpose the data-base format SPINE has been developed at CPM. This format is used by the CPM companies both in their internal data-bases and in the data-base at CPM where the companies share LCA data with each other.

The Swedish case studies presented in this subsection are examples of LCA studies in companies which are members of CPM. These companies have been testing and using LCA in a number of applications, and they are all working with finding ways to work with LCA in a routine manner. The use of LCA is quite broad, and ranges from product and process development to support for market communication. The Ericsson, Perstorp and Akzo Nobel case studies can be said to loosely correspond to these three types of applications respectively. However, some general differences can be observed among the CPM companies. Industry can be classified in two categories, process- and product industries according to what they use LCA for. The manufacturing industries, e.g. large manufacturers of complex products such as mechanical companies, use LCA in their product development. In order to do this they need a single parameter value of environmental impact from the product. Therefore are valuation indices are important to this category of companies in order to chose which product has the least environmental impact. The process industries such as chemical and paper companies use LCA for process development. They are not helped by indices; instead they analyse the inventory results directly to find hot spots in order to trim their production processes. A third category are the LCA beginners who use LCA to visualise the environmental performance of their company (Tillman 1998).

5.2.4.2 Akzo Nobel Surface Chemistry

The company

Akzo Nobel was created in 1994 by the merger of the Dutch and Swedish chemical companies, Akzo and Nobel Industries. Through the merger, Berol Nobel became Akzo Nobel Surface Chemistry. This is a sub-business unit within the Chemicals business unit which produces bulk and speciality chemicals. Akzo Nobel is an international company with more than 70 000 employees in 50 countries and its headquarters in Arnhem in the Netherlands. The turnover in 1995 was ten billion ECU: In its turn, Akzo Nobel Surface Chemistry has its head-quarters in Stenungsund, Sweden, and has 20 plants/sites world-wide where tensides, flotation agents for the minerals' mining industry, and so on, are produced.

The merger did not result in any substantial changes in the company's environmental policy, nor in its dedication to the Responsible Care programme of the chemical industry.

The case-study

The studied LCA project began in 1993 at Berol Nobel, and ended in 1994 when it became part of Akzo Nobel Surface Chemistry. For the project, the "viscose niche" at Berol Nobel/Akzo Nobel Surface Chemistry collaborated with the "non-woven group" at the Institute for Fibres and Polymers (IFP). The viscose niche is a major producer of process chemicals for viscose fibre production. The object of the study was to compare viscose and polypropylene fibres.

A clear start of the LCA project is difficult to identify. LCA as such had been discussed for some time among the non-woven group members who were producers of either viscose or synthetic fibres. The viscose niche manager at Berol Nobel was very curious about LCA and took part in these discussions as a member of the non-woven group. He also met separately with the IFP director to discuss the possibility of IFP to conduct an LCA for Berol Nobel. Various project ideas were discussed. However, afterwards it became a non-woven group project. The reason put forward was that the person intended as LCA analyst at IFP was the programme leader of the non-woven group and he was funded by all members of the group. As a consequence, the LCA study ought to involve all members and not just Berol Nobel. This settled the object of study: a comparison of viscose (a wood-based fibre) with a synthetic fibre was relevant to all group members.

Two other factors were also important. At the time, the non-woven group members had heard about an LCA study conducted at a European branch institute, but they were unable to get hold of it. In addition, the viscose niche was the organiser of an international conference series for the viscose industry, and the viscose niche manager wanted the LCA project to be presented at the next conference. He hoped that this would inject some optimism into the viscose industry which since long was fraught with a bad environmental reputation and was losing market shares to the synthetic fibres. In contrast, he was convinced that the viscose fibre had a future thanks to it being based on a renewable resource (wood).

The viscose niche manager discovered after some time that the project was not moving ahead. His opinion was that IFP did not have the competence to carry out the study, but he had also found an LCA analyst within his own company and he saw to getting him involved in the study. This LCA analyst was someone who had just finished his MSc Diploma project (an LCA study) and was at the time employed on a short-term contract for another LCA project. Although a beginner, he was more experienced than IFP's LCA analyst.

The manager of the LCA analyst at Berol Nobel was an important person although he mostly stayed in the background. He felt that traditional environmental work often "got stuck" and that LCA had something new to offer. One of his strategies was to let the young LCA analyst become established on his own merits. He also gave the viscose niche manager permission to use the LCA analyst as much as necessary.

When the two LCA analysts first met, they started to concretise the project: the two fibres had to be studied in a product application so that a functional unit could be defined. The product chosen was the cover tissue in nappies. Polypropylene had 96% of this market, viscose the rest. A detailed LCA study and proper LCA methodology ("all the way to valuation") were aims in themselves. A familiar valuation method was chosen: EPS, a valuation method based on people's willingness-to-pay to avoid environmental impact.

The two LCA analysts met approximately once a month. Their data collection turned out to become a trade of data for results with many of the prospective data suppliers. The Akzo Nobel LCA analyst always tried to develop a more personal relationship with data suppliers or to use established channels, such as those of the purchasing department with many of Akzo Nobel's suppliers. The idea of using established channels came from his manager. The thought was that established channels would provide a normal and undisturbing way of collecting data.

The total time spent by Akzo Nobel's LCA analyst was approximately 20% of his working hours for half a year. Estimated on the basis of data collection activities, the Akzo Nobel LCA analyst did most of the work.

The results were unexpected: With EPS, it was shown that viscose was better than the market leading polypropylene[18]. This led to a heated debate in the project. At an emergency telephone conference, the members of the non-woven group discussed how the results were to be presented at the conference. It was decided to "scrap" the EPS results in order to get acceptance for the results. Instead, it was agreed to step back from the evaluation and only go as far as the inventory and characterisation results. This was a way of avoiding discussions about evaluation methods when one could have a discussion of the results. The agitated discussions never led to changes in the calculations, only to other formats for result presentation. The manuscript of the speech was faxed to all non-woven group members for approval. The LCA analysts got their clearance the day before the conference.

After the conference, the Akzo Nobel LCA analyst quickly put together a report. He saw it as important to have something to show to the various stakeholders. He expanded the analysis of the results by adding a dominance analysis of the viscose life cycle. It showed that a few industrial activities were responsible for the main part of the total environmental impact and that the contribution from Akzo Nobel's own production was small. Not only did he present the results internally at Akzo Nobel Surface Chemistry (i.e. to the informal LCA committee), he also visited data suppliers. On one of these visits, he showed the data supplier that the NO_x emissions from their pulp mill were smaller than those from their forest transportation. His impression was that "It was something they had never ever thought of before". Later, this data supplier started a logistics project to reduce transportation emissions. Akzo Nobel Surface Chemistry started a similar project which led to a switch from lorries to railroad on certain distances. In fact, the LCA analyst frequently used the viscose example in his presentations afterwards. Especially the dominance analysis was considered effective in communicating the usefulness of LCA. The LCA analyst was eventually given 25 minutes

[18] The reason for this is that EPS places much weight on the use of fossil fuels.

to present his work to the members of the Chemicals' Board. Two weeks later he received full employment as LCA specialist and the existence of the LCA committee was formalised.

The project was generally considered successful, and it led to several new LCA projects. For example, the satisfied viscose niche manager commissioned an update of the viscose results a couple of years later, but that is another story.

Noticeable is that the project served many purposes. Looking only at Akzo Nobel, the project was partly driven by research-based curiosity and by inadequacies with experienced other environmental tools (the LCA analyst's manager spoke about the potential of the tool; the viscose niche manager was intrigued by it; an environmental officer at the pulp mill found it "fun" and wanted to learn more). A more formal reason was market communication (inject some optimism into the viscose industry). In addition, the LCA analyst saw it as an opportunity to establish himself and LCA methodology within the company. To sum up, the project led to expected as well as several spin-off outcomes (new LCA projects, transportation projects, a formalised LCA organisation). This indicates that LCA was used for "organisational learning" rather than for decision making.

Present and future use of LCA

Akzo Nobel's LCA analyst played an important role in the company's work with LCA. Although the project was formally ended after the conference, he kept communicating the usefulness of the LCA study. This shows that his was a conscious plan to introduce LCA, one which he shared with his manager. It can said that both worked in an entrepreneurial manner. Other examples of their entrepreneurial strategies are the conscious data collection strategies and how they worked to obtain a mandate from management. Their own explanation as to why they could act so freely is that there is a "decentralisation culture" in the company.

Since the ending of this particular project, LCA activities have become further established and integrated in the activities of the company. Three years later (1997/98), a second person was employed to work with LCA at Akzo Nobel Surface Chemistry and the LCA committee was reformed into the "Environmental Strategy Group".

5.2.4.3 Ericsson

The company

Ericsson is one of Sweden's largest international corporations. In 1996, it had 94 000 employees worldwide, of which 44 000 were in Sweden. Ericsson which has its business in telecommunications was in 1996 divided in five business areas:

- Radio,
- public telecom,

- business networks,
- components and
- microwave systems.

Privatisation of the telecommunications markets throughout the world caused considerable changes in Ericsson's market situation. From having had a market with only a few hundred customers, the company's new market now consisted of several million customers. The company is also characterised by its rapid growth: its turnover has increased 5-fold over the last ten years.

In its environmental policy, a general commitment to Life Cyle Thinking is made (Ericsson 1996).

The case study

In 1996, a product specialist with part-time environmental responsibilities at Ericsson Business Communication (EBC), which was a part of Business Networks, initiated an LCA project. The project concerned a comparison of alternative media for product documentation for one of their products, a business switching exchange. The documentation describes how the switching exchange is set up, operated, constructed and integrated in the national telecommunications network.

A chance meeting by the coffee machine in the communal coffee room with a documentation specialist triggered the project. The documentation specialist enthusiastically told the product specialist about an environmental discussion at their last group meeting. The reason for the environmental discussion was that the documentation group were looking for new arguments. At the time, the documentation was printed on paper and consisted of more than 60 000 pages collected in some 80 ring folders. They had been working for a long time to modernise the documentation system, to change from paper-based information to electronic-based information (e.g. on disk, CD, www), without success. The documentation manager realised that such a change would dramatically reduce the use of paper, hence the reference to new, environmental arguments. Technical and economic arguments had earlier failed to persuade the product management to decide about the future of the documentation system. The documentation specialist described 'documentation' as an area overlooked by the product managers - it was considered to be something trivial and in no need of big investigations or formal decisions.

The product specialist was on the lookout for ideas for LCA projects. In 1994, he had become convinced that LCA was "the only technique to look at a *product's* environmental impact"; it was product-oriented which suited him as a product specialist, and it was quantitative which he thought would appeal to the many engineers at Ericsson. He even believed that LCA was the only environmental method Ericsson should build environmental work on. After the coffee, the product specialist thought that he should make use of the situation; it had been the first time that someone had approached him to talk spontaneously about environmental matters. He therefore started to plan an LCA project to promote a modernisation of the documentation system.

The planned LCA study was not the first one within the corporation although it was the first within this particular business area. The product specialist had thought that the company's earlier LCAs did not live up to proper LCA practice, and he wanted to use the project as an opportunity to introduce proper LCA methodology (according to ISO 14040) to the company. To obtain proper LCA, he wanted to cooperate with CPM[19]. Another idea was that a particular software should be used. The product specialist's plans for "large scale" implementation of LCA was based on the use of this particular software which could be used for LCA calculations as well as building up an internal corporate-wide LCA database. To sum up, the LCA study was conducted at the department of the product specialist, not at the documentation department.

An LCA analyst (a Diploma student) was recruited via CPM. He spent approximately two weeks studying the LCA concept which was new to him. The ISO standard was abandoned at an early stage since the LCA analyst did not find it instructive. Instead, reports of other LCA studies were studied. Then, around two months were spent on data collection. Finally, around two months were spent on making the calculations with the requested software. There were two reasons why it took so long. Since product documentation is not a core product, there were no strong ties to the data suppliers. Also, several bugs were identified in the software. This led to three new upgraded versions of the software during the time of the project. The LCA as such was a "full LCA". Three valuation methods (EPS, the eco-scarcity method and the environmental themes method) were used for comparison's sake[20].

The Diploma student was to present the results of the LCA study at a seminar at EBC. The product specialist wanted the seminar to be an opportunity to market LCA. To start with, he sent out invitations not only to various people within EBC which was the tradition but also to people throughout the corporation. Given that the interest for environmental issues was low at Ericsson and that the company did not really have any environmental problems, the product specialist had to convince people about the need for LCA in spite of this. There was even a view within the corporation that their products were part of the solution to environmental problems: *"Our technology is in itself adapted to the environment..."* (Ericsson 1996). His strategy was to not to contest the general opinion. Instead of setting out to prove that the company had environmental problems he said that the study was to prove what people already knew. He also referred to the environmental questions from "the market".

The LCA study showed that a CD-based documentation system is preferable to a paper-based one (the approximate relative environmental impact between the two alternatives was 1:1 000). Through the LCA, the documentation people ac-

[19] See also subsection 5.2.4.1 and Box A.
[20] See Bengtsson (1998) for the description of the methods.

quired a technical overview of the documentation system which thus far had been missing, which led to a discussion of choices of different types of CDs and transportation. The study also put the documentation system on the agenda of the product managers. The product specialist, eager to show results of the environmental work, got the project presented in the annual corporate environmental report of 1996, although the project was not finished until 1997. The product specialist went over to work full-time on environmental issues during the LCA project. His promotion of LCA activities in a corporate-wide context made his nearest manager react. The LCA seminar gave the manager the opportunity to call a meeting with the corporate environmental manager and the business area's environmental manager to discuss the division of tasks, responsibilities and costs related to LCA activities between business areas and central corporate units.

Present and future use of LCA

Two reasons lay behind the project. From the perspective of the documentation people, internal political reasons were at the fore: to find new arguments for modernising the product documentation system. From the perspective of the product specialist, the launching of LCA was at the fore. The outcomes of the project were manifold: increased knowledge, discussion about areas of responsibility, networking, etc. This represents a general organisational learning dealing with documentation systems as well as with environmental issues.

Ericsson had made some kind of general commitment to LCA (e.g. in its environmental policy and through membership in CPM). A couple of LCA studies had previously been conducted within various business areas. However, LCA activities were not co-ordinated. This is in contrast to the activities of the product specialist. The LCA study on documentation systems would never have taken place if it had not been for the product specialist. Not only did he initiate LCA activities within his own business area, but also within the corporation as a whole. An example of this is the project to set up a corporate-wide LCA database and the corporate-wide invitation to the seminar. The product specialist thought he had to push the LCA activities forward very much on his own, without much help from management. In other words, the product specialist can be described as an LCA entrepreneur who tested "all" ways to introduce LCA. The way the LCA study was introduced at the seminar is an example of the LCA entrepreneur's implementation strategy and how it was situationally adapted.

5.2.4.4 Perstorp Flooring

The company

Perstorp Flooring is an affiliate of the Perstorp Corporation, which was founded in 1881. The founder family still controls the corporation. The corporation has throughout its history been focused on formalin production and its use in various applications e.g. in decorative laminate, which is one of the Perstorp's most famous products.

Perstorp Flooring is the inventor of laminate flooring with extensive distribution in some 30 countries in Europe, North America and Asia/Pacific. The brand Pergo® is, since its introduction in 1989, the most sold laminated flooring product for residential areas in the world. The production of flooring takes place in Trelleborg, Sweden and in Raleigh, N.C., USA and the flooring laminates are produced in Perstorp, Sweden. The turnover of Perstorp Flooring exceeds 310 million ECU and there are over 1 100 employees.

The case-study

The case study goes back to an LCA of the laminate material in the flooring made as a Master of Science Project. This LCA study of laminate flooring at Perstorp Flooring was initiated due to the increased environmental awareness in the building material sector. In 1994, the R&D project manager tried to find someone who could help Perstorp make an LCA of the laminate flooring. The Chalmers IndustriTeknik (CIT) accepted to do the study and one PhD student working there became responsible for the project. The involvement of the PhD student affected the study somewhat since he wanted to test the Nordic Guidelines on LCA (Lindfors 1995) and to try some allocation principles. The objectives of the study were to:

- Learn more about the environmental impacts of laminate flooring. This aim included identifying hot spots in the life cycle and getting advised on environmental improvements.
- Get independent documentation on the environmental loading of laminate flooring.
- Construct a database that could be used in subsequent LCAs and in product development. This aim included learning how to use the software LCA inventory tool.
- Possibly use the LCA results in future comparisons with other LCA results on competing floor types.

The processes contributing more than four percentage to the environmental load, according to any of the three weighting methods, namely EPS, environmental themes and eco-scarcity, were considered "hot spots". The "hot spot" list below comes from Sjöberg et al. (1997); the processes are listed without ranking order:

- The land use of forestry,
- production of raw materials for the resins,
- production of substrate,
- production of TiO_2 for décor paper,
- waste management of used floor: incineration and land filling and
- indirect impact of energy recovery at waste incineration.

An improvement analysis with possible changes to improve environmental performance was also made with the help of the list of hot-spots. Perstorp Flooring made changes to the hot-spots which they could control.

Present and future use of LCA

LCAs will become more important in the future as stakeholders request more and more environmental information. As mentioned before, there are several stakeholder groups that drive the company to improve their environmental work (Årsredovisning 1996). Perstorp Flooring will continue to elaborate their environmental declarations and provide their customers with the information they need.

Perstorp Flooring only makes one product Pergo® flooring, which is made in several models. All models are very similar in composition and therefore only small modifications of the already contributed study are necessary. Routines to do these updates of the study continuously as modifications are made to the products have been establishment. In the future one major change to the data-base would be the introduction of the user phase of the floor, in order to get more information on a more complete life cycle.

Perstorp Flooring does not see any connections between their LCA and governmental activities. One of the drivers for the environmental work has been certification according to various environmental certificates e.g. EMAS, BS 7750, ISO 14001 and the Nordic eco-label, Green Swan. The LCA study has been a great help to get these certificates but not a prerequisite. The lack of regulations within the field of LCA is perceived as giving freedom to shape the LCA study in order to gain the information most valuable for the company.

In the near future, Perstorp Flooring will improve their data-base in order to update and expand less the processes. Perstorp Flooring also plans to modify the weighting methods to make them consistent with the environmental priorities of the company. Perstorp Flooring believes that the LCA tool will become more effective in the environmental management practice of the company if these changes are made.

Box A: Multi-stakeholder collaborative approaches on LCA

Companies using LCA for the first times are often confronted with complex methodological and practical issues. Despite some codification by the ISO norms 14040 and 14041 (which are limited to the general rules and the inventory phase), several of these issues are still valid, particularly for SME's, which have limited human and financial resources to dedicate to LCA activities. The main problems are usually the availability and quality of data from outside the company, the allocation rules, the impact assessment and the interpretation phase.

Therefore, companies might profit a lot from exchanging their experience with each other, learning from already existing examples and tackling with unsolved methodological issues in a collaborative approach. Moreover, if the latter includes not only business but also public authorities and research institutes and universities, a general collaborative working framework is created, which might significantly support the development and diffusion of LCA. In particular, this multistakeholder approach might be very helpful for creating and fostering a collaborative attitude between business and legislators (instead of confrontation) about the correct use of LCA.

This is indeed what happened in *Sweden* with the establishment of a national centre of excellence in LCA and product-related environmental assessment, *Centre for environmental assessment of prod-uct and material systems (CPM)* in 1996. CPM is jointly funded by industry, The Swedish National Board for Industrial and Technical Development (NUTEK), and Chalmers University of Technology. The overall goals of CPM are is to build up a long-term framework of knowledge that is relevant to the needs of industry in the field of product-related environmental assessment. The member companies of CPM are: ABB AB, Akzo AB, Avesta Sheffield, Cementa AB, Duni AB, Electrolux AB, Telefon AB LM Ericsson, Mo & Domsjö AB, SCA Mölnlycke AB, Norsk Hydro ASA, Perstorp AB, Stora Corporate Research AB, Vattenfall AB, AB Volvo. A major instrument in reaching this goal is the establishment of an LCA data-base. For this purpose the data-base format SPINE has been developed at CPM. This format is used by the CPM companies both in their internal data-bases and in the data-base at CPM where the companies share LCA data with each other.

The Swedish case studies presented in this subsection are examples of LCA studies in companies which are members of CPM. These companies have been testing and using LCA in a number of applications, and they are all working with finding ways to work with LCA in a routine manner.

The creation of a similar working framework is the main goal of the *Italian Association for LCA*, which was officially founded in early 1999. Its establishment was particularly supported by a common initiative of business with some Italian universities. However, the association includes, beyond companies and universities, also public authorities and public companies, other research institutes, consultants, NGO's and consumer organisations. Most of Italian companies using LCA are participating to the Association. To date, among our case-studies, the list includes FIAT, ABB Italy and Italtel.

While it is too early to assess the actual impact of this Association on the diffusion of LCA in Italy, there are high expectations about its potential role. The association is meant to be a forum of discussion on LCA issues and experience. Several informal working groups were created in 1998 before its formal establishment. To date, the main topics discussed by the working groups are general methodological issues, an observatory on international experiences, energy systems, simplification rules and life-cycle costing.

5.3 Overview on application patterns

We carried out 20 case-studies with the intention of looking for the application of LCA within business. Whereas in section 5.2, the findings of 12 selected case studies were presented in a descriptive way, we have given an overview on the situation of LCA within all 20 companies examined by us in this section.

In order to systematise the information, the findings are presented in four tables combining different characteristics for each of the 20 companies as follows:

1. The *company*: This part delivers some technical information (turnover, size, product groups, environmental image, competition, main customers, type of products).
2. The *LCA-study* which we considered:
   - What was examined? Points addressed are theme and year of LCA-study.
   - Who joined the process? Points addressed are actors (initiators, commissioners, promoters, supporters and opponents), LCA-resources and mandate/involvement of top management.
   - Motives: Points addressed are goals, objectives and motivations.
   - How was the study carried out? Points addressed are type of LCA, considered LCA-steps and life-cycle and single product or comparison among different products.
   - Outcomes: Points addressed are surprising results, dissemination of results of the LCA-study, additional spin-off effects and level of satisfaction.
   - Role of LCA.
3. A single LCA-study does not describe the applications of LCA. In part III, we describe the *LCA-activities* of the examined companies as of today, namely:
   - Influence of LCA within the company: Points addressed are the role of LCA within environmental management system, product innovation and marketing.
   - Future: Points addressed are expansion of LCA-applications, use of LCA as a routine management support tool, expansion of range of products and future diffusion into other areas of the company.
   - Policy: Points addressed are the role of policy as a pushing factor for LCA and expectation of business towards the role of policy within LCA.
   - LCA-status: Points addressed are the start of LCA-activities, present areas of LCA-applications, frequency of LCA, LCA-software, data-base, Know-how/capacities, formal organisation for LCA and the status of LCA according to the degree of institutionalisation.

The 20 case-study companies are presented country by country. Within each table, the order of the presentation of the companies follows the stage of institutionalisation of LCA which has been observed by us (increasing use from the left to the right)[21]. The entries in the fields of the tables are based on our case-studies.

---

[21] See section 5.4 for an exhaustive analysis of the institutionalisation process.

5.3.1 Switzerland

Table 5.2. Overview on application of LCA within five Swiss companies

Aspects	Holderbank Cement and Beton (HCB)	Landis & Gyr Utilities (LGUT)	Ernst Schweizer AG	Ciba Speciality Chemicals (CSC)	Kraft-Jacobs-Suchard (KJS)
I. The company:					
Headquarters	Siggenthal (CH)	Zug (CH)	Hedingen (CH)	Basel (CH)	Bremen (D)
Turnover (in ECU)	400 mill. (1996)	625 mill. (1996, group: 4 500 mill.)	40 mill.	625 mill. (1996, whole group: 4 000 mill.)	3 500 mill. (1997)
Size	2 000 (1996)	5,300 (1996, whole group: 31 000)	400	4 000 (1996, whole group: 20 000)	4 300 (1997)
Product groups	Cement, concrete, other building materials	Counters for electricity, services and systems for energy efficiency.	Windows, doors, solar energy systems mainly with metals	Pigments	Branded food products (Coffee, cream cheese, confectionery)
Type of competition	Price	Quality	Quality	High added-value products and technologies	Quality
Type of customers	Construction	Energy-suppliers, consumers	Construction	Industry, consumers	Retail trade
Type of main products	Simple mass products, commodities	Electromechanical products, services	Specialised products	Commodities	Commodities
II. The specific LCA-Case-study:					
What?					
Object of the LCA-study	Cement and concrete	Counter for electricity	Post-box and windows for winter gardens	Pigment Irgazin DPP Red BO	Roasted coffee incl. packaging
Year of LCA-study	1997/98	1992	1995	1993-95	1995-96

Aspects	Holderbank Cement and Beton (HCB)	Landis & Gyr Utilities (LGUT)	Ernst Schweizer AG	Ciba Speciality Chemicals (CSC)	Kraft-Jacobs-Suchard (KJS)
Who?					
Initiators	Top management	Environmental manager	Environmental manager	Division's environment, health and safety group	Internal project group (members of environmental department)
Commissioner	Association of the Swiss Cement-, Lime- and Plaster Manufacturers	Top management of Division	Top manager	Top management of division	Decision of top management
Promoters ("entrepreneurs")	Environmental manager	Environmental manager	Environmental manager	EHS manager	Department public affairs
Resources (LCA-analysts)	Internal/ External	External	External	Internal, some external support	Strong external, modest internal
Other involved departments					
• Supporters	Working group (representatives from production, third parties)	None	None	Production sector	KJS R&D (Munich)
• Opponents	n.a.	n.a.	n.a.	n.a.	n.a.
Mandate / involvement of top management?	Yes / no involvement	Yes / no involvement	Yes / high involvement	Yes / no involvement	Yes / modest
Why?					
Goals	To support the production optimisation process, comparison with other building materials	Analysis of the different phases of the LCA, which one is most important	Comparison of existing products with hypothetical new products and evaluation of highest impact of a product	Determination of (ecological) weak points and methods for improvements	Determination of impacts, best and worst case, effect of various ways of action

Aspects	Holderbank Cement and Beton (HCB)	Landis & Gyr Utilities (LGUT)	Ernst Schweizer AG	Ciba Speciality Chemicals (CSC)	Kraft-Jacobs-Suchard (KJS)
Objective at the beginning	Retrospective (consecutive eco-controlling)	Retrospective	Prospective	Retrospective	Mainly retrospective
Motivations	Internal/ external	Internal	External	Internal/ external	External (environmental discussion, decree on packaging)/internal (integration of environmental aspects)
How?					
Type of LCA	Quantified	Quantified	Quantified	Quantified and simplified	Quantified
LCA-steps	Complete (interest focused on inventory)	Complete	Complete	Inventory, impact assessment, interpretation	Complete
Product's life-cycle	Complete	Complete	Complete	Complete	Complete
Single product or comparison?	Single product (comparison of alternative fuels and raw materials substitutes)	Single product	Comparison	Single product (comparison of two production sites)	Single product (comparison of scenarios regarding growing and processing)
Outcomes					
Surprising results?	No (LCA not yet accomplished)	Yes, use-phase is the most important phase	No, existing product was the best	No	No
Internal dissemination of results	Not yet	Information to designers and marketing; publication of result in eco-efficiency-report.	Yes	Yes: article in company journal, presentation to the other divisions	Yes: presentation to Philipp Morris (parent company)

Aspects	Holderbank Cement and Beton (HCB)	Landis & Gyr Utilities (LGUT)	Ernst Schweizer AG	Ciba Speciality Chemicals (CSC)	Kraft-Jacobs-Suchard (KJS)
Additional spin-off effects?	Not yet	No	Yes	No	No
Satisfaction?	n.a.	Yes (good base for Product Eco-Performance [PEP])	Yes	No, disillusionment because LCA had almost no consequences	High ("Success model coffee-LCA")
Role of LCA					
• Learning	Modest (ecological) problems are already known)	High	High	Modest	High
• Integration of LCA to support specific decisions	Not yet	Integration of results of LCA in PEP	Yes	No	Yes
Actions as consequence of learning/decisions:					
• Changes in organisation	No	No	No	No	No
• Product development	Not yet	Yes (integration of criteria in PEP)	No (not necessary)	No (no direct connection between LCA and measures taken)	Yes (test of new transport packaging material, study on production energy saving measures)
• Marketing	Not yet	Modest (short communication in Eco-Efficiency Report)	Modest	No	No
• Other external communication	Not yet	no	Yes (in environmental report)	No	No

Aspects	Holderbank Cement and Beton (HCB)	Landis & Gyr Utilities (LGUT)	Ernst Schweizer AG	Ciba Speciality Chemicals (CSC)	Kraft-Jacobs-Suchard (KJS)
• Other actions	Not yet	No	No	The LCA has triggered an energy saving project	Start of co-operation project "best practice development for growing and processing of coffee beans"
III. LCA-activities in general as of today:					
Influence of LCA in the company					
Environmental management system / objectives	Not yet	Yes	Yes	No	Yes but modest
Product innovation	Modest	High	Modest	None	Modest
Marketing	Use of LCA not excluded in the future	No use of LCA	No use of LCA	No use of LCA	No use of LCA
Future					
Future expansion of applications?	Scarce	Scarce	Scarce	Scarce	Scarce
Future routine management support tool?	Yes (as eco-controlling instrument)	Scarce	No	Scarce	Yes (optional and supplementary tool)
Future expansion of range of products?	n.a.	Yes, if data base for electronic goods will be available	Yes	Scarce	A last comprehensive study on the third main product category is planned
Diffusion into other organisation (corporation) areas?	All areas covered	Scarce	No	Scarce	No
Policy					
Policy: role as pushing factor	Low (discussion on energy taxes)	Low	Medium	Low	Low (decrees on packaging)

Aspects	Holderbank Cement and Beton (HCB)	Landis & Gyr Utilities (LGUT)	Ernst Schweizer AG	Ciba Speciality Chemicals (CSC)	Kraft-Jacobs-Suchard (KJS)
Expectations of business towards policy	None	Development simpler methods and data base	Development of methods and data bases	None	Opposition against "generic" use of LCA by authorities (legislation)
LCA-Status					
Start of LCA-activities	1997	1992	1994	1992	Ca. 1990
Present areas of application	(see above)	Two studies available	Two studies	Two studies underway	No routines
Frequency	First	Some	Some	Some (only one covering the entire life-cycle)	Some
LCA-software	External	External	External	Internal	External
Data-base	Mainly internal	Internal/ external	External/Internal	Internal/ external	External / internal
Know-how/capacities	LCA-know-how with Staff Office Environment, internal capacity very small	know-how mainly by consultant, internal know how meanwhile available	External and partly internal	Know-how within pigments and consumer-care divisions, capacity within pigments division small	High internal know-how, capacity modest
Formal organisation for LCA?	No	Project management by environmental manager	Environmental manager	No, project management by environment, health and safety staff resp. product safety	No
Status	Habitualisation stage	Habitualisation stage	Habitualisation stage	Objectification stage, but standstill	Objectification stage, but decreasing importance

5.3.2 Germany

Table 5.3. Overview on application of LCA within five German companies

Aspects	Weleda AG	Ytong AG	AEG Hausgeräte GmbH	Bosch and Siemens Hausgeräte GmbH	Henkel KGaA
I. The company:					
Headquarters	D for the company, owned by Weleda Switzerland	D for the whole Ytong holding owned by Readymix Concrete (UK)	D for the company, but since 1994 owned by Elektrolux (S)	D, joint venture between Siemens AG (D) and Robert Bosch GmbH (D)	D
Turnover (in ECU) (1996)	60 mill.	440 mill.	1.3 bill.	4.5 bill.	8.2 bill.
Size (number of employees)	500	3 000	7 750	31 000	47 000
Product groups	Pharmaceuticals, cosmetics	Building materials	Electric household appliances	Electric household appliances	Chemicals, metal chemistry, adhesives, cosmetics, washing/ cleansing agents, hygiene products
Type of main products	Commodities	Commodities	Complex final products	Complex final products	Commodities
Type of competition	Quality	Quality and price	Quality (and price)	Quality and price	Price
Type of customers	Retailers, final consumers	Construction industry and private housebuilders (via retailers)	Retailers, final consumers	Retailers, final consumers	Retailers, final consumers
II. The specific LCA-Case-study:					
What?					
Object of the LCA-study	Packaging	Building blocks	Weights of washing machines	Soap containers	Surfactants
Year of LCA-study	1995	1993	1995	1995	1990-92

Aspects	Weleda AG	Ytong AG	AEG Hausgeräte GmbH	Bosch and Siemens Hausgeräte GmbH	Henkel KGaA
Who?					
Initiators	Marketing/Logistics	Head of marketing and sales department	Product development of washing machines/ dishwashers/ dryers	Product development of laundry products together with central environmental dep.	Marketing
Commissioner	Decision of the management	Meeting of R&D department and managing board	Decision was taken in the product team of the above product division	n.a.	Management of the washing & cleansing agents division
Promoters ("entrepreneurs")	Logistics	Marketing/ sales (after the completion of the LCA the R&D department overtook this role)	Central technical department	Laundry product area, especially the environmental department	Department Quality and environment of the washing and cleansing agent division
Resources (LCA-analysts)	Modest internal, strong external	External	Modest internal, strong external	Modest internal, strong external	Internal, but some external support
Other involved departments					
• Supporters	Environmental officer, procurement, (in the end the Weleda companies in France and Switzerland co-operated) .	Several YTONG production works co-operated by providing data	Product team with representatives from development, controlling, marketing/ distribution, and production.	Central and group environmental departments, product development of laundry group.	Working group with representatives of other product divisions, marketing, R&D and natural scientists
• Opponents	Weleda companies in France and Switzerland (in the beginning)	R&D, parent company	None	n.a.	n.a.
Mandate / involvement of top management?	Yes / modest	Yes	Yes / modest	Yes / modest	Yes / modest

Aspects	Weleda AG	Ytong AG	AEG Hausgeräte GmbH	Bosch and Siemens Hausgeräte GmbH	Henkel KGaA
Why?					
Goals	Information of consumers Packaging problem Packaging Ordinance	Information of market	Optimisation of weak points German waste policy Activities of competitor	Support decision	Public Relations/ marketing Clarity with regard to surfactants Review of decision
Objective at the beginning	Retrospective	Retrospective	Prospective	Prospective	Retrospective, but accompanying decision
Motivations	Internal / external	External	Internal / external	Internal	External
How?					
Type of LCA	Quantified	Quantified	Quantified and simplified	Quantified	Quantified
LCA-steps	Complete	Complete	Inventory, interpretation	Complete	Inventory, interpretation
Product's life-cycle	Exclusion of the stage of use	Complete	Exclusion of the stage of use	Complete	Production of the surfactants and the proceeding processes
Single product or comparison?	Comparison	Comparison	Comparison	Comparison	Comparison
Outcomes					
Surprising results?	Yes	Yes, especially for R&D and marketing/ sales	No	No	No

Aspects	Weleda AG	Ytong AG	AEG Hausgeräte GmbH	Bosch and Siemens Hausgeräte GmbH	Henkel KGaA
Internal dissemination of results	Yes, namely at a meeting with directly affected departments, management and colleagues from Switzerland and France	Yes, namely at a meeting with sales/ marketing staff and at a meeting with production department	Yes, namely at a meeting with product team, environmental department, central technical department	Yes, namely at a meeting with management and environmental departments Written information to involved departments	Yes, namely by co-operation with a working group and by presentation of results to management
Additional spin-off effects?	Yes, stimulation of product innovation and encouragement of application in CH and F	Yes, R&D are interested in the tool	No	Yes, namely for working group "Environmentally sound product development"	No
Satisfaction?	n.a.	Sales/marketing not, R&D yes	High	High	High
Role of LCA					
• Learning	Very high	Very high	Very high	High	Very high
• Integration of LCA to support specific decisions	Yes	No	Yes	Yes	Yes
Actions as consequence of learning/decisions:					
• Changes in organisation	No	No	No	No	No
• Product development	No, but stimulation	Some	No, but stimulation	No, but encouragement of planned decision	Yes
• Marketing	No	Some	No	No	No

Aspects	Weleda AG	Ytong AG	AEG Hausgeräte GmbH	Bosch and Siemens Hausgeräte GmbH	Henkel KGaA
• Other external communication	Only for interested persons	Whole study published; information for interested persons	Only for highly-interested persons and researchers	Only for interested persons and researchers	Only for interested persons and researchers
• Other actions	Yes, procurement	None	None	None	None
III. LCA-activities in general as of today:					
Influence of LCA in the company					
Environmental management system/ objectives	Yes	No	Yes, but modest	Yes, but modest	Yes, but modest
Product innovation	None, but indirect stimulation	Modest	Important	Modest	Important
Marketing	Yes	No	No	No	No
Future					
Future expansion of applications?	Yes	No	Yes	Yes	Yes
Future routine management support tool?	No	No	Occasionally	No, only for weak-points and important decisions	Yes, but as supplementary tool
Future expansion of range of products?	Yes, but modest	No	Yes, but case by case	Yes	Yes
Diffusion into other organisation (corporation) areas?	Yes, but in the future	No	Yes, but controversial	Modest	Yes
Policy					
Policy: role as pushing factor	Strong	Low	Medium	Low	Low
Expectations of business towards policy	Should increase with specific supporting measures	None	Should increase with specific supporting measures	Should increase with specific supporting measures	Should increase with specific supporting measures

Aspects	Weleda AG	Ytong AG	AEG Hausgeräte GmbH	Bosch and Siemens Hausgeräte GmbH	Henkel KGaA
LCA-Status					
Start of LCA-activities	1994	1993	1992	Beginning of 1990s	1990
Present areas of application	No routines	No routines	LCA is used at check-points for product development	No routines	LCA might take part within routines for important product decisions
Frequency	The first	Some (updating the existing ones)	Numerous	Some	Numerous
LCA-software	No	No	Yes, for testing	Yes	Yes
Data-base	No	External	No	External	Internal
Know-how/capacities	No	Low; LCA-capacities are not existing	Yes, but modest	Yes	Yes and own consultancy within the group
Formal organisation for LCA?	Not existing	Not existing	Not existing	n.a.	Existing
Status	Habitualisation stage	Habitualisation stage	Objectification stage	Objectification stage	Sedimentation stage

n.a. information not available

5.3.3 Italy

Table 5.4. Overview on application of LCA within five Italian companies

Aspects	GlaxoWellcome Italy	Cartiera Favini	ABB Italy	Italtel*	Fiat Auto
I. The company:					
Headquarters	Great Britain	Italy	Sweden	Italy	Italy
Turnover (in ECU) (1996)	619 million for the two Italian plants	41 mill. for the whole Favini Group	1 805 mill. for ABB Italy (27 bill. for the whole Group)	2 011 mill.	40 176 mill. (for the whole Group)
Size (number of employees)	2 025 employees for the two Italian plants	221 employees in the Favini Group 134 employees in Cartiera Favini	210 000 for the whole ABB group. Over 2 200 employees between the 3 SACE plants	15 892 employees overall; 2 100 employees in L'Aquila's site	238 000 employees (whole Group)
Product groups	Pharmaceutical products	Paper	Power generation, transmission & distribution systems and devices	Telecommunication systems and networks	Cars
Type of main products	Complex products	One simple product	Complex products and commodities	Complex products and commodities	Complex products
Type of competition	Long tradition & innovation	Quality, innovation and environmental image	Innovation & quality	Long tradition, still mostly related to a protected market position	Price & long tradition
Type of customers	Final consumers and hospitals	Final consumers and paper products manufacture industries	Public sector or industry	Public and private telecommunication industries	Final consumers

Aspects	GlaxoWellcome Italy	Cartiera Favini	ABB Italy	Italtel*	Fiat Auto
II. The specific LCA-Case-study:					
What?					
Object of the LCA-study	Two alternative packaging systems for ampoules (secondary package produced internally)	Raw materials used in paper production	Low voltage circuit breaker	Telephone packaging	Engine block materials
Year of LCA-study	1997	1995	1996	1995	1995/96
Who?					
Initiators	Environmental management at the group level	Top management	Top management at the group level	Eco-compatibility and new technologies function (at the central engineering department)	Head of Department of Environmental & Industrial Policies (DAPI)
Commissioner	H.S.E. (Health, Safety and Environment) management at the national level	Top management	Director of ABB Ricerca (Italian ABB R&D centre), top management of ABB Sace	Central Engineering department	Head of DAPI
Promoters ("entrepreneurs")	H.S.E. management at the national level and product development group (packaging technologists)	Internal environmental group (different functions involved)	LCA appointed function at ABB Ricerca	Eco-compatibility and new technologies function	Head of DAPI and LCA analyst
Resources (LCA-analysts)	Internal: product development group	External support (consultants)	Internal to the group (One person at ABB Ricerca – study coordinator) and 5 people internal to ABB SACE (coord. by the Product Manager)	Internal to the group (Eco-compatibility) and external (Research Insistute)	Internal; one full-time person at the Central Laboratories of DAPI Restricted involvement of external consultants for learning purposes

Aspects	GlaxoWellcome Italy	Cartiera Favini	ABB Italy	Italtel*	Fiat Auto
Other involved departments					
• Supporters	Loss prevention management	No	ABB Italy Country Environmental Controller (CEC)	Quality department	Research Centre of FIAT Group
• Opponents	None	None	None	None	None
	Mandate / involvement of top management?	Yes	Yes / Involvement in the environmental group	Yes / top management both of ABB Sace and ABB Ricerca chose the object of LCA	Yes / decision to start LCA
Why?					
Goals	Find possible environmental improvements in the packaging area	Find environmental improvements in paper's raw material supply	Test of the LCA methodology	Test of the LCA methodology	Evaluate environmental performances of two engine block's material
Objective at the beginning	Retrospective	Retrospective	Mainly retrospective	Retrospective / mainly prospective*	Prospective
Motivations	Corporation	External	Corporation	Internal. / Internal *	Internal
How?					
Type of LCA	Quantitative	Quantitative	Quantitative	Quantitative / Streamlined*	Quantitative
LCA-steps	Goals and scope, inventory (as of the time of the case study)	Inventory and interpretation (partly impact assessment)	Complete	Complete / n.a.*	Complete
Product's life-cycle	Gate to gate (production- as of the time of the case study)	Cradle to gate (extraction and production stages of raw materials)	Complete	Complete / gate to gate*	Complete
Single product or comparison?	Comparison	Single product	Comparison of new vs. old product	Single product	Comparison of new vs. actual engine block

Aspects	GlaxoWellcome Italy	Cartiera Favini	ABB Italy	Italtel*	Fiat Auto
Outcomes					
Surprising results?	Partly: the inventory quantified unexpected material and resources intensity of the alternative packaging	Yes	Partly (from the impact assessment / interpretation phase)	In part / in part*	Yes
Internal dissemination of results	Only within involved functions (at the time the case study was carried out)	Partly in the environmental report (distributed to all employees)	Yes within both Italian and international corporate level meetings	Publication of the LCA but no specific internal dissemination / formal communication of both project and results to all interested functions within the Company	Yes, to top management and Technical, Purchasing, and Production departments
Additional spin-off effects?	Weak-spots identification of production	No	Weak-spots identification, such as resources and energy saving potentials	No	No
Satisfaction?	Yes	Yes	High and willingness to pursue an LCA strategy within all Italian ABB companies	LCA is considered too complex, although the methodology still needs additional research and standardisation*	High, increasing willingness to integrate LCA in the product innovation process
Role of LCA					
• Learning	High	Low (no development of internal capacities for LCA)	High	Yes / Yes*	High

Aspects	GlaxoWellcome Italy	Cartiera Favini	ABB Italy	Italtel*	Fiat Auto
• Integration of LCA to support specific decisions	n.a.	Yes , LCA results have contributed to encourage new R & D areas for raw material supply	Yes but no direct consequences on product in this case	No / High *	Yes
Actions as consequence of learning/decisions:					
• Changes in organisation	No	No	Yes (appointment of a new function for LCA)	No / Yes*	Yes (enlargement of the LCA team)
• Product development	No	Yes (for raw material supplies)	No	No / Yes*	Yes
• Marketing	No	No	No (LCA is very shortly mentioned in the technical brochure of the product)	No / No*	No
• Other external communication	No	Yes, in conjunction with other environmental communication initiatives (articles, environmental report, conferences..)	Within specific meetings	Yes (publication of LCA) / No*	To experts within specific meetings (i.e. SAE, SETAC); also briefly mentioned in the environmental report
• Other actions					
<u>III. LCA-activities in general as of today:</u>					
Influence of LCA in the company					
Environmental management system /objectives	No	No	No	No / No*	No
Product innovation	Premature at this time	Partly, further R & D on alternative raw materials for paper	Not at the moment, but strong willingness to integrate LCA results in product innovation	No / Yes as soon as the methodology is able to be fully integrated*	Yes, within the forthcoming "Integrated Development Plan of Components" (IDPC)

Aspects	GlaxoWellcome Italy	Cartiera Favini	ABB Italy	Italtel*	Fiat Auto
Marketing	Premature	No	LCA is not regarded as a marketing tool	No / No*	LCA is not regarded as a marketing tool (yet)
Future					
Future expansion of applications?	Premature	No	Yes	Yes / Yes*	Yes
Future routine management support tool?	Likely	No	Yes	No / Yes *	Yes, within IDPC
Future expansion of range of products?	Yes	No	Yes	Yes / Yes*	Yes, full LCA on new components, within IDPC; screening LCA on whole cars
Diffusion into other organisation (corporation) areas?	Yes, probably marketing	No	No	No / No*	Yes, within IDPC
Policy					
Policy: role as pushing factor	Low	Medium	Low	Medium / Low*	Low
Expectations of business towards policy	Quite uncertain	Medium (green label)	Low / Medium	Low / Low*	Medium (pro-active position with respect to the EU end-of-life vehicles directive)
LCA-Status					
Start of LCA-activities	1997	1995	1996 in ABB Italy	1995	1994
Present areas of application	Learning	Stopped	Studies are being carried out in other ABB Italy companies	Learning / increasingly used for product development*	Used within design of components (towards a systematic use within IDPC)
Frequency	2 LCAs by 1998	1 LCA (stopped)	Some	3 LCAs by 1998	Numerous

Aspects	GlaxoWellcome Italy	Cartiera Favini	ABB Italy	Italtel*	Fiat Auto
LCA-software	External	No	Internal (modified version of software used at international corporate level)	External	Internal (external software used for comparison / learning)
Data-base	Internal (at the time of the case study)	External	Internal	External and internal	Internal
Know-how/capacities	Developing internal capacities	External	Internal to the ABB Group	Italtel is building its own internal LCA capacities	Internal, significant
Formal organisation for LCA?	No, the LCA will integrate in the existing N.P.I.T. (New Product Introduction Teams)	No	Yes, by appointment of an LCA reference person in each ABB company	No	Yes, LCA team at Central Laboratories at DAPI, formal routines for internal communication
Status	Habitualisation stage (carrying out new LCAs in areas where it is applicable, such as packaging and devices).	Habitualisation stage; first LCA; the company has not expressed the willingness to start new LCAs.	Objectification stage Introducing LCA with strong support from top management and the goal to make it a routine tool)	At the border between habitualisation and objectification stage for LCA. Objectification for the simplified tool (carrying out new LCAs studies for main types of products , while progressing in the alternative methodology for implementation in product innovation processes)	Sedimentation stage

* Items indicated with * refer to the specific "Eco-compatibility methodology" developed by Italtel
n.a. information not available

5.3.4 Sweden

Table 5.5. Overview on application of LCA within five Swedish companies

Aspects	Ericsson	Perstorp Flooring	Akzo Nobel Surface Chemistry	Norsk Hydro	Volvo[a]
I. The company:					
Headquarters	Sweden	Perstorp (Sweden)	Sweden (with corporate headquarters in NL)	Oslo (Norway)	Gothenburg (Sweden)
Turnover (in ECU) (1996)	14 billion	1.5 billion	558 million	10 billion	17.2 billion
Size (number of employees)	94 000	7 000	70 000	38 000	70 333
Product groups	Telecommunication	Laminate flooring	Surfactants	Refinement of natural resources for food production, energy and materials	Automobiles
Type of main products	Complex	Simple, commodities	Commodities	Simple	Complex
Type of competition	Quality	Quality	Quality & price	A change over from price to function is ongoing	Quality
Type of customers	Industrial clients, retailers	Retailers and building companies	Industrial clients	Retailers, but a change to approach the final consumers is ongoing	Retailers, end consumers
II. The specific LCA-Case-study:					
What?					
Object of the LCA-study	Product documentation system for switching exchange	Laminate flooring	Viscose & polypropylene fibres	Cleaning processes for the dairy-industry	Technical concepts/car
Year of LCA-study	1996-97	1996	1993-94	1997-98	1996-97

Aspects	Ericsson	Perstorp Flooring	Akzo Nobel Surface Chemistry	Norsk Hydro	Volvo[a]
Who?					
Initiators	Product specialist within Business Communication business area	Environmental officer	Viscose niche manager & non-woven group at IFP (branch institute for fibres & polymers)	Industrial chemistry division	Conf.
Commissioner	As above	As above	Viscose niche manager	Initially internal research centre within Norsk Hydro. Later in the project a joint venture between Norsk Hydro and Norwegian dairy association	Conf.
Promoters ("entrepreneurs")	As above	Environmental officer and environmental consultant	LCA analyst	LCA-analyst at Norsk Hydro research centre and researchers at Tine	Conf.
Resources (LCA-analysts)	External	Both internal and external. (Mostly external)	Internal & external	Joint venture by the involved parties	Internal
Other involved departments					
• Supporters	Documentation department, print shop, various individuals	Research department at the surface materials division.	Purchasing department, environmental officer, LCA steering committee at Akzo Nobel Surface Chemistry; non-woven group at IFP	Industrial chemistry division; research centre of Norsk Hydro; Hydro Agri Europe; Tine Norske Meierier; Henkel-Ecolab	Central LCA group
• Opponents	Existing, but not identified	No opponents of the study are known.	None identified	No opponents of the study are known	Conf.

Aspects	Ericsson	Perstorp Flooring	Akzo Nobel Surface Chemistry	Norsk Hydro	Volvo[a]
Mandate / involvement of top management?	Yes (business area managers), partial funding of project	Yes/ modest	No	Yes, but not in this particular LCA.	Conf.
Why?					
Goals	Support modernisation of product documentation system; Introduce LCA	Learning	Compare viscose and polypropylene fibres; Present results at international viscose conference	The LCA is supposed to be used for an environmental declaration of the product	Conf.
Objective at the beginning	n.a.	Retrospective	n.a.	Retrospective	Conf.
Motivations	Internal & corporation	Internal	Cooperation	Internal and external)	Internal
How?					
Type of LCA	Quantitative	Quantitative	Quantitative	Quantitative	Quantitative
LCA-steps	Complete	Complete	Complete	Complete	Complete
Product's life-cycle	Complete	Complete except from user phase	Complete	Complete	Complete
Single product or comparison?	Comparison	Single product	Comparison	Comparison	Comparison
Outcomes					
Surprising results?	Yes, to some extent	Yes	Yes	The project was not finished within the study time of the project.	Conf.
Internal dissemination of results	Yes, namely presentation for representatives from various parts of the corporation	Yes selected parts have been presented to involved parties. The complete report has	Yes, namely presentations for viscose niche and boards of managers	The project was not finished within the study time of the project.	Yes, namely presentation for employees within the car corporation

Aspects	Ericsson	Perstorp Flooring	Akzo Nobel Surface Chemistry	Norsk Hydro	Volvo[a]
		not been spread (to not reveal the production technology.	internal report was written		
Additional spin-off effects?	Yes, i.a. a clearer division of responsibilities of LCA work and a better overview of the documentation technology.	New production methods, change of the used materials in the flooring, conference paper and plans for a Perstorp evaluation method. Information source for work with ISO 14001, EMAS and BS 7750 certificates.	Yes, i.a. a transportation project, and new LCA projects (see below)	The project was not finished within the study time of the project.	Conf.
Satisfaction?	Modest	High	Yes	See above	Conf.
Role of LCA					
• Learning	Yes	Yes	Yes	Yes	Yes
• Integration of LCA to support specific decisions	No	No	No	No	Yes
Actions as consequence of learning/decisions:					
• Changes in organisation	Full-time environmental responsibility for product specialist	Yes, personnel has been assigned to do the continuos updating of the study	Employment of LCA analyst & formalisation of LCA steering committee	The project was not finished within the study time of the project.	Conf.
• Product development	No	Yes	No	See above	Conf.
• Marketing	No	Yes, implicit through environmental certificates used in mar-	Long-term strengthening of viscose market	Intended to be used as information to product declarations.	Conf.

Aspects	Ericsson	Perstorp Flooring	Akzo Nobel Surface Chemistry	Norsk Hydro	Volvo[a]
		keting			
• Other external communication	Project reported in corporate environmental report; Presented at seminar for Swedish industry on IT & the environment	Yes, conference paper	Conference presentations, presentations to suppliers	The project was not finished within the study time of the project.	Conf.
• Other actions	Simplified version used in staff training	-	Transportation project, new LCA projects	-	Conf.
III. LCA-activities in general as of today:					
Influence of LCA in the company					
Environmental management system / objectives	Yes	Yes	Yes	Yes	Yes
Product innovation	Visionary	Yes, modest	Indirect stimulation	Modest	Yes, some
Marketing	Yes	No	Yes	Yes	Yes
Future					
Future expansion of applications?	Yes	Yes	Yes	Yes	Yes
Future routine management support tool?	Yes	Yes	Yes	Yes	Yes
Future expansion of range of products?	Yes	Only one product	Yes	Yes	Yes
Diffusion into other organisation (corporation) areas?	Yes	Yes	Yes	Yes	Yes
Policy					
Policy: role as pushing factor	Low	Low	Low	Low	Low
Expectations of business towards policy	Low	Low	Low	Low	Unclear

Aspects	Ericsson	Perstorp Flooring	Akzo Nobel Surface Chemistry	Norsk Hydro	Volvo[a]
LCA-Status					
Start of LCA-activities	1994	1994	1991	Late eighties	1989
Present areas of application	Selected product areas visionary concepts "the future office".	Flooring materials	All main product areas	All main product areas	Production processes, components, car
Frequency	Occasionally	Continuos updating of the study	Numerous	LCA department working with LCA matters.	Numerous
LCA-software	Yes	Yes	Yes	Yes	Yes
Data-base	Yes	Yes	Yes	Yes	Yes
Know-how/capacities	Yes, modest	Yet limited, but improving	Yes, expanding	Large	Yes
Formal organisation for LCA?	Under construction	To come	Yes	Yes	Yes
Status	Habitualisation stage; introducing the tool	Between habitualisation and objectification stage	Objectification stage	Objectification stage	Sedimentation stage

a The company declared that the information on the specific LCA-study which we considered was confidential. Therefore, we have not been allowed to present some information on Volvo.

Conf. Confidential

n.a. Information not available

5.4 Conclusions

In this subsection we summarise the results from the 20 case-studies identifying common patterns between different companies; we have analysed the case-studies within the analytical framework of the "Institutionalisation Theory". This theory aims at describing the introduction of an innovation in a company, from the early stage in which the new idea was generated to the late stage in which the new approach or tool is fully integrated within daily business activities.

By applying this analytical framework to the introduction of LCA in companies, we were able to interpret all 20 case-studies and to identify the main factors which may lead LCA respectively to success or to failure in a company.

This subsection is further subdivided into four items. The first subsection 5.4.1 introduces the institutionalisation theory with specific respect to LCA. The second one (subsection 5.4.2) explores in more detail the role of LCA within this context. Subsection 5.4.3 describes common patterns and classifies the 20 analysed companies accordingly. Finally, subsection 5.4.4 summarises the main factors which determine the successful introduction and application of LCA within companies.

5.4.1 From learning to doing: the institutionalisation theory

The case studies describe LCA projects in companies with varying degrees of experience of working with LCA: Some are beginners, others have been working with LCA for some years. Together, all cases depict a process of introducing LCA as a regular tool in the companies' environmental practice. Such a process is called "institutionalisation" using the terms of organisation theory. An overview of institutionalisation theory has been elaborated by Tolbert/Zucker (1996). What is interesting about institutionalisation theory is that it describes the characteristics of different phases in the introduction of a new phenomenon until that new phenomenon becomes something taken for granted. It also describes some key factors important for full institutionalisation. Therefore, we have used the theory of institutionalisation and "translated" it so that it describes the process of LCA introduction. Examples from our case studies are used to illustrate various aspects of this process.

The theory envisages three stages of the institutionalisation process, after the new idea of an innovation has been defined:

Innovation → Habitualisation → Objectification → Sedimentation

The *habitualisation* stage is the first stage of innovation application within the company. Often it concerns a small part or a restricted area of the company (e.g. the environmental department in the case of LCA). The next stage, namely the one during which the new idea or tool begins to be more widely used within the company is called the *objectification* stage. This is very likely the most crucial phase of the whole process. It is usually at this stage that the future adoption of the innovative idea or tool is determined. If the innovation is further systematically integrated within business activities, one enters the final stage of the institutionalisation process, called the *sedimentation* stage. Table 5.6 summarises the main characteristics and the factors influencing the different stages of the institutionalisation process. The latter is described in more detail in the following paragraphs.

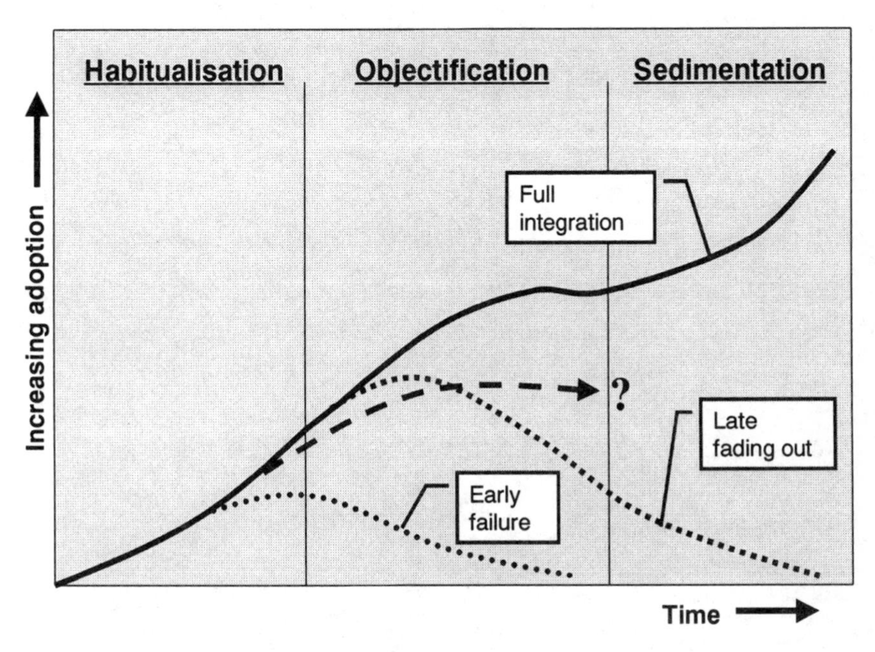

Fig. 5.1. The dynamics of institutionalisation of LCA

The different institutionalisation stages correspond to a different level of adoption/integration of the innovation and/or the innovative tool (LCA in our case) within the company. Figure 5.1 shows the level of adoption of an innovation within a company in function of time. In principle, four possible trajectories of adoption pattern are possible: The upper line in the figure shows an adoption curve leading to the full integration of the innovation within the company. The intermittent line represents the case of "uncertain companies", in which the "destiny" of the innovation is still unclear. In these companies there are both positive indications, which suggest for a further integration of the innovation, as well as some negative signals which indicate that there might be a failure in the integration process. The two other possible (negative) adoption paths (dotted in figure) are an early failure and the late fading out of the innovation. As already mentioned earlier and shown in the figure, the crucial phase is the the one of semi-institutionalisation called objectification in the theory, during which the "destiny" of the innovation is most likely determined.

Table 5.6. General stages of institutionalisation and comparative dimensions (adapted from Tolbert/Zucker 1996)

Dimension	**Pre-institutionalisation stage: Habitualisation**	**Semi-institutionalisation stage: Objectification**	**Full-institutionalisation stage: Sedimentation**
Characteristics of adopters	• Few, similar circumstances • Homogeneous	• Heterogeneous	• Heterogeneous
Impetus for diffusion	Imitative	Imitative/ normative	Normative
Theorisation activity	None	High (from problem to solution)	Low
Variance in implementation	High	Decline	Low
Structure failure rate	High	Moderate	Low
Consensus	No consensus	Some kind of consensus	Consensus
Main characteristics	• Generation of new structural arrangements in response to an organisational problem and formalisation of arrangements in procedure/policies of the organisation • Creation of new structures in organisation is largely an independent activity • Structure tend to be impermanent	• Diffusion of a new structure • Definition of an organisational problem • Development of a general and shared attitude to problem solving behaviour • Structures tend to be more permanent, more widespread	• Existence of a formal structure spread across actors: Structure is the institutionalisation • Survival across generations of members of organisation/ high maintenance of action over time • High transmission of action • High resistance to change
Important factors for this stage	Adoption is determined by: • Internal political arrangements • Technical factors • Economic factors	Adoption is determined by: • Champion (entrepreneur) with two tasks, a) identify organisational problem b) justification of a particular formal structure as a solution to the problem • Monitoring of competitors and efforts to enhance relative competitiveness	Adoption is determined by: • Importance and resistance of opponents • Cultural support/ promotion by advocacy groups • Demonstrable results • Low costs

5.4.1.1 Habitualisation

Just the fact that LCA has been invented and exists is not enough for companies to start working with LCA. For companies to adopt LCA, the people in the companies must feel a need for LCA. This takes place during the first stage of the institutionalisation process, the habitualisation stage, when signals for when to apply LCA are identified. Examples of such signals could be the awareness of external factors, such as the environmental debate and/or scientific discussion, or the presence of a specific problem along the life-cycle (typical example: waste). Other examples are pressure from market such as public procurement and/or consumers' pressure.

In general, at this early stage of the institutionalisation process, technical and economic factors together with internal political arrangements largely predict adoption. On the whole, there are generally few adopters at this stage. Since organisational decision makers may share a common core of knowledge and ideas, the adoption of a given innovation may, and often does, occur in close association with other organisations. In Sweden, a group of ten companies formed the Product Ecology Project in 1991 in order to learn and develop business-adapted ways of working with LCA. In Switzerland there has been rather close co-operation between industry and research institutions. Moreover, LCA activities in this country have been supported and accompanied by a "Green" industrial association, namely ÖBU. In Germany, some companies have carried out LCA studies together (i.e. AEG with Daimler Benz). In Italy, most LCA companies are still in the habitualisation stage; however, a strong collaborative attitude is observed, mainly in connection with the forthcoming Italian Association for LCA.

Imitation may follow from such association, but this is not necessarily so since there is no consensus on the general utility of LCA at this stage. This means that early adopters of LCA are/were companies in somewhat similar circumstances but without similar ways of working with LCA. Examples of external technical/economic factors triggering LCA activities are for instance the reaction to environmental debate (e.g. at Fiat) and/or inquiries from the market. Marketing turned out to be a very important driver for LCA particularly in Germany. This is confirmed in our case-studies in the two examples of AEG and Weleda, which started their LCA activities also with the goal of reaching/improving their eco-image on the market. In some cases the external triggering factor might come from legislation. However, in our cases this was confirmed only with respect to the regulation of packaging and waste in Germany (example: Weleda, AEG and KJS) and to public procurement in Switzerland (example: Ernst Schweizer).

Regarding internal political arrangements, the picture is mixed. In Sweden, LCA is often introduced in a bottom-up approach. Environmental officers or engineers start working with LCA due to dissatisfaction with other environmental tools and/or due to research-based curiosity. It is not until after the initial steps that they work to obtain the support and a mandate from their management to continue with these LCA activities. In contrast, in Germany, Switzerland and Italy, LCA is often introduced in a top-down approach. In Italy, the case-studies also confirmed the relevance of corporate groups as the driver for starting LCA (e.g. at ABB Italy and Glaxo-Wellcome Italy) as shown in the survey (see also subsection 4.2.1.4).

Knowledge among non-adopters of what constitutes an LCA study is typically very small at this early stage. In other words, those not directly involved in LCA activities have probably never heard about LCA. This is mainly due to the lack of horizontal communication and insufficient vertical communication within the company. Very rarely specific meetings and/or activities are organised (the biggest exception are the "ad-hoc" cross-division working groups set-up at Henkel from the very beginning).

Apart from the decision to take up LCA, some kind of organised ways of working with LCA are also starting to develop. However, such structures are typically short-lived. These early structures for working with LCA are typically unique from company to company. An example of how short-lived these early "LCA organisations" are comes from the Akzo Nobel case study: a reference group which was set up to support a specific LCA project. This rather informal reference group lasted only approximately six months and reformed itself to become an LCA steering committee. This change marks the transition into the next stage of institutionalisation, i.e. objectification. Of course, such organisations are short-lived in cases where LCA was ultimately stopped, e.g. in the Swiss company Ernst Schweizer, which had the typical issue of limited resources of an SME.

5.4.1.2 Objectification

For the LCA implementation process to move on, a consensus regarding the use of LCA has to develop after the initial adoption of LCA. The development of such a consensus concerning the use of LCA is probably the most crucial stage before LCA is used routinely, and is the main characteristic of the objectification stage. It involves the development of a general and shared understanding of LCA and the use of LCA. This is necessary for the use of LCA to expand beyond its point of origin. Consensus can emerge through two different mechanisms:

One mechanism works through monitoring and gathering information about LCA activities in a company's business surroundings. Monitoring is a way for a company to confirm its adoption of LCA or to assess the risk of adoption. The larger the number of companies working with LCA, the lower the hurdle for adoption. The Ericsson case study provides an example of how this can be used. At the presentation of the results of his first LCA project, the project leader thought it was important to also present CPM[22] to the invited colleagues. The purpose was to show that even if the LCA activities were new and unknown to most people at Ericsson, such activities were not unusual in Swedish industry. The im-

[22] See Box A.

plicit logic was that if "all other" companies are working with LCA, why should not Ericsson do it as well?

Monitoring activities can be very direct, as in the case of Henkel vs. its competitor Procter & Gamble. In this case, monitoring (and marketing) was also a direct driver for starting LCA activities at Henkel. However, monitoring activities can be more indirect, as in the case of Fiat Auto, which participated in the collaborative sector study of EUCAR together with many other European car manufacturers. This is also the case of ABB Italy, which ensured monitoring through its corporate group (the electro-mechanical sector is very active with respect to LCA activities).

The *second* mechanism works through a champion who can promote LCA[23]. A champion for LCA is generally a person with a strong environmental background and sensitivity. This is both true in the case of a top-down approach (i.e. the head of DAPI at Fiat Auto), and it also holds in the case of bottom-up introduction (e.g. the R&D officer at Ernst Schweizer). Moreover, in the case of bottom-up approach, she/he might typically be someone with a material stake in the promotion, be it a permanent job as in the case of Akzo Nobel, Henkel or BSH.

In any case, whatever the type of introduction of LCA in the company (either top-down or bottom-up), in order to be successful, the champion needs to achieve two tasks of theorisation:

- to identify a consistent pattern of dissatisfaction in the organisation thereby diagnosing a generic organisational problem to which the use of LCA can be a solution.
- to justify the use of LCA as a solution to the diagnosed problem on logical or empirical grounds.

This means that the champion needs to be persuasive, mainly by providing examples of successful and useful LCA studies. In the cases of Akzo Nobel and Ericsson, the champions had similar strategies for launching LCA. An example of this strategy is "go where they open the door for you", i.e. the champion was attentive to what interest others showed and conducted only studies where there was an interest, instead of conducting the most "appropriate" LCA. Apart from this, there are large differences between these two cases, and the champions make situational adaptations of their LCA promotion. At Ericsson, where there was little recognition of their environmental problems, the champion did not go against the general view. Instead he justified the use of LCA by referring to inquires from customers and to the idea that LCA can be used to prove what everybody "already" knows. At Akzo Nobel, however, the champion who could refer to the well-known environmental problems and the poor reputation of the chemical industry provoked the older colleagues by saying that old thinking had led to to-

[23] In the following, we use both the name of champion or entrepreneur. See also Box B.

day's problem and that new thinking (LCA and young people like him) was necessary for finding solutions[24].

Whatever the strategy adopted, the existence of a valid justification for LCA at this particular stage is a crucial factor for a successful diffusion of LCA within the company. We observed that in all cases where good justification was missing, either LCA activities slowed down or stopped, or they had an uncertain future, or alternative methodologies were developed (see also subsections 5.4.3.2-4)

Adopters in general during this stage promote LCA and at the same time monitor the accumulation of evidence of its utility. The use of LCA during this stage has a somewhat fashionable quality. In other words, technical/economic/political characteristics are no longer as important for predicting adoption, and adopters represent a more heterogeneous group. The structures built up for working with LCA are still relatively short-lived. They may last perhaps a couple of years as in the case of the LCA steering committee at Akzo Nobel before it was transformed into the Strategic environmental committee.

5.4.1.3 Sedimentation

Sedimentation is characterised by the cessation both of the promotion of LCA and the accumulation of evidence for an LCA. By now, there is a formal structure for working with LCA. This means that the use of LCA has become taken for granted, routine, institutionalised, and survives even if the LCA-people in the company change. The higher the level of institutionalisation, the more easily structures are transferred to new people. An example is given at FIAT Auto, where the coordinator of the LCA-team has remained the same, but several other LCA-analysts have changed during the last two years.

Obstacles that can hinder institutionalisation are of two kinds:

- Either there is a lack of demonstrable results,
- or there is a group of actors adversely affected by the use of LCA who collectively mobilise against it.

No really organised opponents were identified in any of our case-studies. However, in some companies there are various people who say that the environmental trend is just a fashion and a waste of time (and money).

---

[24] A more comprehensive review of LCA projects at Akzo Nobel and Ericsson in the light of institutionalisation theory is found in Baumann (1998).

Box B: Actors involved in the institutionalisation process of LCA

We identified three crucial actors within the institutionalisation process, namely the initiator, the entrepreneur (or champion) and the set of champions. Their relevance changes along the process phases. The *initiator* is the central driver in the first phase. The importance of the *entrepreneur* (who might be, but is not necessarily, the same person as the initiator) is central in the objectification phase. His/her presence and influence are crucial factors for the overall success of the whole process. He/She is the one who develops a strategy for creating/emphasising a good justification for LCA and reaching a consensus within the company. Finally, in the sedimentation phase, when large consensus is needed, the *set of champions* involved in the diffusion of the tool has to expand within different sectors of the company. The set of champions generally includes the entrepreneur, however this is not strictly necessary, since in the full sedimentation phase the tool "survives" even if the people change.

Other actors involved in the LCA activities are the *commissioner* (the one who really gives the money) and the *LCA analyst(s)*, who physically carry out the study. However, their role is less important, as long as they do not coincide with the initiator and/or entrepreneur.

5.4.2 The role of LCA

5.4.2.1 A changing role with time

By shifting the focus from the overall LCA activities to the particular LCA studies, the use of LCA in decision making can be discussed. A description of its use has to combine an examination of the LCA study's role (be it legitimising, learning or something else) with an identification of the area of application (be it product development, purchasing, or other).

One main result of our case-studies is that the role of LCA changes with time, according to the different institutionalisation stages. For instance, all LCA studies in the habitualisation stage were carried out for learning and confirmation purposes. This rather logical observation is true even in the case of Henkel, in which high expectations and a great deal of resources were connected with LCA activities from the very beginning.

In the next, the objectification stage, we observed that the role of LCA (in the "successful cases"[25]) changes, slowly shifting from a retrospective to a more prospective use. This means that LCA is not only used for supporting decisions already taken, but also to assess the environmental impacts of alternative solutions (new materials, components or entire products) in advance.

---

[25] Successful means that application of LCA continues in a specific company.

Finally, we observed that the role of LCAs in the sedimentation stage can be different from case to case, but it is always very precise. The reason for these differences may have to do with simplification: for example, at Fiat the complete LCA-method is now systematically applied to the choice of materials and components, but only screening LCAs are applied to the whole car. Similarly, at Henkel an LCA is now used in an "incremental" way for every "important" product innovation, based on new combinations of (well known) chemicals.

In particular, the clearest example of the changing role of LCAs we observed in our case-studies refers to marketing. In several cases, in particular in Germany, marketing was an early driver for starting LCA studies[26]. Quite soon, all firms had to admit that, given the current methodological issues, LCA cannot yet be directly used for marketing purposes[27].

In some cases, this was an important reason for stopping/decreasing LCA activities (e.g. the Swiss company Ernst Schweizer and the German company Ytong). However, in many other cases this did not prevent the use of LCA for other, more internal purposes (R& D, product development, design, etc.). This was also further supported by the many spin-offs of LCA. The typical comment we heard several times in connection with this was "Even if the results are too complex and sometimes disputable to be used for marketing, nevertheless LCA revealed many weak-spots and improvement opportunities. We still think that it is the correct tool for assessing the environmental properties of product and to make alternative choices in the early design phases".

This changing role is strictly connected to the possible learning processes within the company, and also to the reaction of people with respect to unexpected results. We found that this rather subjective[28] factor is a very crucial one for the "success" of LCA within the firm (see also the next paragraphs). Obtaining satisfaction with an LCA is not only related to the LCA delivering the expected outcomes but also to whether the LCA produces surprising results and whether the actors have the ability to make use of these unexpected insights.

---

[26] In particular, there were big expectations, that LCA (eventually integrated with weighted aggregate environmental indicators) could be used to demonstrate that a product of the company A is environmentally better than the corrisponding product of the competitor company B.

[27] This situation might change in the future, if standard procedures for simplification, assumptions, system boundary definition and external presentation are well defined and codified. Anyway, both the survey and the case-studies show that LCA is used for some kind of information of customers and stakeholders. In particular, Swedish companies use LCA as a basis for environmental product declarations. Whether these are to be included in marketing initiatives or in other forms of external communication is a subtle question.

[28] It is the *perceived* degree of satisfaction about the fulfillment of expectations and the *perceived* possibility for change which do matter.

5.4.2.2 Routine or unique situations?

As it is quite difficult to describe what decision making actually is, there are many different theoretical schools. A number of different ways of using LCAs in decision making were discussed in section 2.1. A differentiation between operative and strategic decisions is often made. The former are based on standard operating procedures (SOPs) and an organisation's memory. Among our case studies, only at Henkel could it be observed that LCAs were conducted within an SOP. Other companies, such as, Volvo and Fiat, have conducted many LCAs, but so far none of these were conducted on the basis of an SOPs [29].

Instead, various circumstances had triggered these different studies. This indicates that the role of LCAs is more to provide new answers for unique situations, which represents a more strategic role for LCAs. In other words, in the absence of LCA-related SOPs, it seems that LCAs are used as instruments for learning in decisions which are uncertain and might lead to changes in products and organisational processes.

5.4.2.3 LCA for learning

LCA comes to serve several purposes. In the habitualisation stage, only a few LCAs were intended to lend support to decisions which had been planned beforehand. One of these is illustrated by the case study of the LCA at BSH. Generally however, the LCA studies had a more exploratory approach. They led to some expected, sometimes also to some unexpected results, and to several follow-up decisions. This is indicative of how LCAs fill a learning role in companies.

Apart from testing the LCA as such (discussed in relation to institutionalisation), the ambition at the start of an LCA study can be to structure a complex, "messy" situation, i.e. to find out "what really is". In the Akzo Nobel case, they wanted compare how the environmental impact of two types of fibre (related to each other). An LCA was used be to sort out the arguments in the packaging debate in the Weleda case. This shows that the goals attached to an LCA study can vary. With the results however, actors involved in the LCAs reacted in two ways: LCA results were either surprising or unsurprising.

An element of surprise is found in many of the case studies. For example, in the Akzo Nobel case, it was surprising to many that the viscose fibre came out better than expected in comparison with the polypropylene fibre. This new "fact" was then used to boost the optimism of the waning viscose industry. Another example from the same case study was how the dominance analysis unexpectedly pointed out that forest transportation was a significant step in the viscose life cycle, something that later led to changing logistics and the mode of transportation.

[29] This situation might change in the future as well. For instance, at Fiat, the systematic use of the LCA within the "Integrated Development Plan of Components" initiated in 1998 might be considered as a SOP.

At Fiat, the LCA study indicated that aluminium engine blocks are environmentally preferable to conventional cast iron engine blocks only in very particular boundary conditions, which do *not* happen to apply to the present Italian market situation. This changed the prevailing perception in the company that lighter motors (and thus lighter cars) must be environmentally better because of lower fuel consumption.

In other cases, the learning experience can be directed more to confirm expected outcomes and frames of reference. Here, the Ericsson case may serve as an example. The LCA study provided the documentation specialists with yet another argument to support a modernisation of the product documentation system. This was used to "force" product managers to put issues about the product documentation on their agenda. Other examples of LCAs confirming and further supporting decisions already taken are at ABB Italy, and the first LCA-studies at Henkel. This was actually the case in most retrospective studies.

As already mentioned earlier, the "biggest" general surprise (in particular in Germany and Switzerland) was the evidence that at present LCA can not be used (yet) for direct marketing purposes[30]. Of course this led to a certain amount of dissatisfaction in all those cases where marketing had been a main driver for LCA. In some cases companies reacted by fading out LCA activities whereas in others this simply caused a shift towards other (mostly internal) uses of LCA.

Indeed, crucial factors behind satisfaction with LCAs depend on whether LCAs deliver the expected outcomes as well as the actors' openness to surprises and their ability to make use of these unexpected insights. The Akzo Nobel case may serve as an example to describe the importance of actors. Through the LCA, unexpectedly large environmental loads from forest transportation were identified. Although these emissions occurred within the realms of another company, the Akzo Nobel LCA analyst communicated this eye-opener to the other company. In addition, the importance of transportation in that particular LCA study was used to trigger their own transportation project at Akzo Nobel Surface Chemistry. At Henkel, the management decided that even if the results were complex (much more so than expected) LCA identified environmental weak-spots and was the right tool for incremental environmental product innovation.

Going through the cases studies, a lack of satisfaction often goes hand in hand with a lack of surprising results and spin-off effects and with expected goals not being met (see especially Swiss cases studies). Whether this is a problem related to the LCA study or the actors is not possible to say based on the case studies. It is possible that an LCA study has not come up with any surprising results or that some of the results which were surprising were considered irrelevant to the company. However, it is our personal feeling that in some cases subjective factors may have prevailed (e.g. at Ciba Speciality Chemicals).

---

[30] We emphasize again that LCA can be and *is used* for information and education of clients and stakeholders however.

5.4.2.4 *From learning to action*

The presence and absence of surprise shows that two types of learning take place. The model of single-loop and double-loop learning developed by Argyris/Schön (1976) describes these two ways of learning (see also section 2.1). Single-loop includes problem solving and assessment of alternatives through the application of knowledge and models, whereas double-loop learning also includes transformational learning which implies a reflective decision maker who through learning may redefine his/her framework. Here, the framework typically consists of the way of looking at the products, production systems, and the relative environmental responsibility of actors at different steps in the life cycle. Worth noticing is that Argyris' and Schön's model indicates that the type of learning is actor and context dependent. Here we make use of this model in connection with LCA.

Learning cycles

LCA sometimes produces unexpected results. Basically, the question is how companies make use of and handle these results. There are various options for reacting to unexpected LCA results:

1. Companies can ignore them,
2. they can re-interpret them as being irrelevant,
3. they can deal with them by extending time horizons and
4. they can actually change their products and/or processes.

The options companies chose depend on the company specific context and can be described in terms of learning situations and learning cycles (Hedberg/Wolff 1999). Logically there exist four learning situations:

1. A situation where the way a company perceives its theory of business as usual LCA-results do nothing to change this, neither the "theory-of-business" nor the actions and strategies that are chosen as a consequence of the theory.
2. The second situation is characterised by applying the same basic "theory-of-business", but the LCA may possibly have an impact on the actions chosen. Products and or processes are then changed as a result of the studies' result.
3. The third situation occurs when the LCA-study actually indicates that changes have to be implemented, but companies' decision makers for various reasons chose not to change their present actions and strategies (materialised in products and processes).
4. The fourth situation, in principle, occurs when the LCA study both implies changes of the "theory-of-business" and these are related into strategies and actions.

In summary the principle decision situations are presented in the following Table 5.7:

Table 5.7. Principle learning situations (Hedberg/Wolff 1999)

	Theory-of Business	Strategy and actions (in terms of processes and products)
1	Same	Same
2	Same	Different
3	Different	Same
4	Different	Different

As LCA's intention is to give more insight and knowledge about the impact of a process or product during its life cycle, the analysis of this impact in terms of learning is a central issue in our project and research. We will now elaborate shortly on the various learning outcomes that we observed.

The clearest example of learning is the shift from the intended use of LCA for marketing purposes towards more internal applications. In many cases[31], particularly in Germany, the start of LCA activities was driven by external demand translated by marketing, or by external demand from stakeholders. However, companies had to realise quite quickly, that at the current state-of-the-art of the methodology, LCA cannot be used for marketing. Of course, this was the first unexpected result. But in many cases there was another unexpected result, namely that LCA triggered alternative methods, R&D, spin-offs, direct product development, collaborative behaviour (development of best practices), etc.

We observed that in companies in which simply the first unexpected result was the main reason of dissatisfaction (LCA could not be used for its intended application, that is marketing), LCA activities faded out. In contrast, where the second unexpected results were fully accepted and "internalised", further LCA activities were developed.

As to the question whether and how LCA results were translated into actions, we observed that:

- In the short term, only those actions which represent a win-win situation were adopted; in contrary situations cost reasons would always prevail.
- In the long term (seeing, not acting) they were generally translated into actions, for instance in the next investment cycle[32].

[31] This is not true in Sweden, where LCA activities were usually initiated internally, in a bottom-up approach, and promoted by environmental and/or R&D officers.

[32] A good example for this is the case of FIAT, which intended to adopt aluminium engine blocks for the next car models (with a specific take-back and recycling system) and not for the present ones.

Organisational context

The use and diffusion of LCA has to be understood in its organisational context. As our studies indicate, the institutionalisation of LCA as a tool for understanding the life cycle of products and processes and also for changing these, depends on the organisational setting for its outcomes.

We now see a tendency in global industries to organise themselves in so called strategic business units (SBU). Also, a focus on processes from production to customers emerges in addition to the SBU structure. We assume therefore, that the process orientation will require two things: an integration of LCA into the chain of decision making in processes and as a consequence, a simplification of LCA methods in order for it to adapt to the speed of decision making in process driven businesses. In fact, we find these two tendencies in our case-studies.

One example for the simplification of LCA methods is the case Landis & Gyr and Italtel, where on the base of LCA-results new tools prepared specially for product designers have been developed (called respectively "Product-Eco-Performance" and "Eco-compatibility analysis"). An example of integration is HCB, where the LCA is intended to be used as an eco-controlling instrument for the whole company.

5.4.3 Common patterns

Based on the above-mentioned considerations of institutionalisation theory and the role of LCA, we have tried to classify the different company-subjects of the case-studies into several groups with similar behaviours and trends. To do this we have used four main classification criteria, namely:

- Trend in the use of LCA (increasing or decreasing),
- Position within the stages of the institutionalisation theory,
- Size of the company,
- Introduction of LCA in the company (top-down or bottom-up).

In addition to these criteria we felt the need to introduce another one related to those companies going to use LCA only in a limited / simplified way in the future. These companies have used LCA in the past but are currently developing alternative methodologies for the assessment of the environmental performances of their products.

Figure 5.2 summarises the classification of the 20 companies of our case-studies according to the above-mentioned criteria.

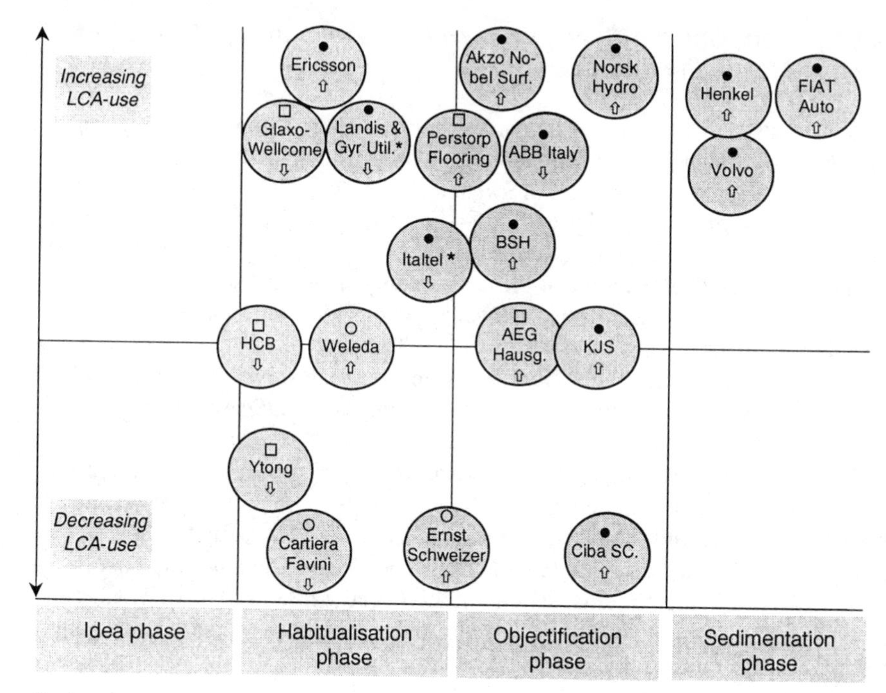

Explanation:
● large companies
□ medium sized companies
○ small companies
⇧ bottom-up-approach in the introduction / habitualisation phase
⇩ top-down-approach in the introduction / habitualisation phase
* companies using a simplified/alternative methodology

Fig. 5.2. Classification of the case-study companies according to the level of institutionalisation of LCA

We were able to identify some common patterns within each of four main groups of companies:

1. Companies which are expected to increase their use of LCA in the future;
2. companies which are developing/applying simolified/alternative methodologies;
3. companies in which the trends of future use of LCA are basically uncertain;
4. companies which are fading out the use of LCA.

5.4.3.1 Companies with an expected increase in the application of LCA in the future

In our opinion, this group of companies includes Henkel and BSH in Germany, Fiat Auto, ABB and Glaxo-Wellcome[33] in Italy, and Volvo, Akzo Nobel, Perstorp, Norsk Hydro and Ericsson in Sweden.

All five Swedish companies and no Swiss firm belong to this group. Although one should avoid misleading generalisations (this result may well be the consequence of the particular choice of the case-studies), nevertheless this is a sign of a generally positive attitude amongst Swedish companies and of a relatively more negative approach amongst Swiss companies. This kind of saturation phenomenon in Switzerland is strongly supported by the result of the survey showing that 23% of (responding) Swiss companies declared that the use of LCA will *not* increase in the future (see subsection 4.2.4.4).

Most companies belonging to this group are large companies. There are only two medium-sized and no small enterprises. This is most probably connected to both the financial and human resources available for LCA activities. In fact we found out that a necessary (although insufficient) condition for the success of LCA in a firm is the "internalisation" of LCA know-how in that firm. No company which still strongly relies on external consultants is expected to increasingly use LCA in the future.

5.4.3.1.1 Companies in the sedimentation stage

We found that only three companies of the 20 analysed case-studies are clearly going to increase the use of LCA in a more or less systematic way and find themselves in the sedimentation stage. The companies are Fiat, Henkel and Volvo.

We identified the following four common crucial factors for success:

- Presence and influence of an entrepreneur,
- good justification for LCA (precise role of LCA, openness of actors with respect to unexpected results),
- existence of good organisation for internal communication,
- large set of champions.

Presence and influence of an entrepreneur

In all cases there is an entrepreneur with a high personal commitment. Even more important, his position is influential within the company. At Fiat Auto, the entrepreneur is the Head of a Department appointed at the very same time for both the Environmental and Industrial Policies of the company. He refers directly to the C.E.O. and the Chairman of Fiat Auto. At Henkel, it is the head of the Quality &

[33] Glaxo-Wellcome Italy and Ericsson are still involved with their very first experiences of LCA. Thus, all speculations about the future use of LCA in these companies should be taken with some care.

Environment Department of the Washing and Cleaning Agents (W&CA) Division. Here, the centrality is guaranteed by the product-oriented internal organisation (W&CA is one of six product-related divisions). At Volvo the situation is slightly different because the introduction of LCA into the company happened much more in a bottom-up approach. However, this does not mean that the entrepreneur is less influential and/or well known. As a matter of fact, the "LCA inventor" at Volvo has been honoured twice since 1988 - first by WWF, then in 1996 by the King of Sweden - for his early ideas on alternative environmental measures in industry, and LCA in particular.

Good justification for LCA

Of course, the presence of an active entrepreneur is not enough. In order to promote the LCA tool, he needs good justification for it. Obviously, the role of LCA changes over time. At the beginning it is mainly for learning, and this becomes more and more precise with increasing experience. In fact, according to theory a precise role of the innovation tool is a main characteristic of the sedimentation stage. For LCA, this is the case of the environmental screening of materials and components at FIAT, and of new compounds at Henkel. At Volvo, the commitment is to make LCA (eventually in simplified forms) a routine working tool for engineers and designers. The goal is simplicity and speed: the environmental material choices made by any construction engineer should not take more than five minutes.

Another justification is given by the monitoring of competitors during the objectification stage. At Fiat, the monitoring was guaranteed by the participation in the EUCAR project. At Henkel, the fact that Procter & Gamble has always been very involved in LCA has been an important motivation for both beginning and continuing LCA activities.

At any stage, the justification has to be supported by positive experiences.

Another very important fact is the openness of actors with regard to unexpected results. In fact, the question whether expectations have been met and results are relevant or not is a rather subjective issue. As explained in the next subchapters, the lack of satisfaction with respect to expectations is one of the main reasons for the failure of the diffusion process of LCA within companies.

In contrast, both at Henkel and Fiat a positive attitude towards unexpected results was observed. At Henkel, management decided that even if the results were complex (much more so than expected) LCA identified environmental weak-spots and was the right tool for incremental (environmental) product innovation. At Fiat, the company changed its former strategy of producing lighter cars with aluminium motors when the LCA demonstrated that this was only positive for the environment under certain conditions[34].

[34] It is worth recalling that this decision was mainly driven by economic reasons. The results of LCA supported and boosted the change of strategy.

Finally, justification can be supported at any stage by a strong company environmental policy. This is the case both Fiat and Henkel, where an early strategy was also motivated by public discussions and concerns about the environmental impact of cars and chemical products.

In the case of Fiat, the environmental policy explicitly declares the importance of Life Cycle Management in the car sector. As an almost straightforward consequence, LCA is considered the right tool for assessing the environmental impacts of materials and car components.

Presence of a good organisation for internal communication

As already mentioned, only at Henkel, LCA is used as a routine tool within Standard Operating Procedures (SOP). However, in all successful cases, a good organisation for the communication of results can be observed anyway. At Henkel, this organisation was formally established from the very beginning. A cross-product division working group of 10-12 people, including marketing, R&D representatives, ecologists and biologists was established. This immediately guaranteed the communication between different departments and divisions. At Fiat, LCA-activities was co-ordinated by the Central Laboratories, which already provided other technical departments with data on materials. Therefore, there was never any lack of confidence: communication was boosted by the fact that LCA data were simply added to other information about materials. The situation will be even more favourable in the future, as LCA will be systematically used within the Integrated Development Plan of car Components. At that point, LCA might even be used at Fiat as a routine tool within SOPs as well.

In both cases, the presence of these "formal" organisations not only improved communication between departments but also facilitated the communication of concise LCA results to management.

Large set of champions

The existence of good communication tools also made the creation of a large set of champions easier. In fact, both the LCA staff and other people from other departments or divisions were involved early in the LCA discussion. The fact that most of these people (mostly belonging to technical departments at Fiat, whereas marketing people were also involved at Henkel), were satisfied with the "tool" LCA, certainly supported its diffusion. We could say that in all cases there was both a top-down (the top management was involved early) and a bottom-up approach.

Other common aspects

Apart from the four above-mentioned crucial success factors, we also identified a set of other important and common supporting factors:

- Presence of a formal LCA organisation/team/staff,
- no strong reliance on external consultants,
- development of comprehensive internal know-how, set-up of internal data-bank and own software,
- mandate from Top-Management (early mandate in the case of Henkel and Fiat, much later in the case of Volvo, where the introduction of LCA happened much more bottom-up),
- early environmental awareness and commitment at the highest firm levels, and
- public discussion on environmental issues in both car and chemical sectors.

Box C: Success factors for the institutionalisation of LCA:
It is worth highlighting that in all of the most "successful LCA stories" among our case-studies we observed interesting connections with the institutionalisation theory. In particular we observed that all the following success factors were satisfied:
- Influence of entrepreneur,
- low status of opponents,
- large set of champions,
- precise application of LCA,
- quite high investments (for establishing internal data-bank and software).

We did not observe any substantial difference between the top-down or bottom-up approach at the beginning of the implementation process, that means that success or failure of LCA-activities do not depend on this criteria.

5.4.3.1.2 Companies in the objectification stage

In our opinion, Akzo Nobel, Perstorp, Norsk Hydro, BSH and ABB Italy all fully belong to this second group of companies. All these companies expressed the intention to expand the use of LCA in the future (this is true at: ABB Italy, Akzo, Perstorp, slightly less at BSH). Therefore we can generally affirm that they are still in the objectification stage and that it is simply a question of time, before they enter the sedimentation stage.

There are several reasons for arriving at this conclusion.

In all companies, there is a very active entrepreneur. She/he might follow different strategies to promote LCA (see subsection 5.4.1), however main justifications for LCA exist in all cases. Monitoring of other companies is also important at this stage (the electromechanical sector is one of the most active in LCA as far as ABB Italy is concerned). Even if LCA was mostly used in a rather retrospective way to confirm already expected results (e.g. ABB Italy, BSH), the level of satisfaction was generally high and a "strategic value" was given to the methodology. In many cases the relevance of LCA was also supported by the fact that it provided several significant spin-offs, such as the identification of weak-points and bottlenecks, (ABB Italy), transportation projects (Akzo Nobel), etc. In several cases it led to a change in organisation and the set-up of a formal LCA function or organisation (ABB Italy, Akzo Nobel).

In all cases this process is supported by a long-term company environmental strategy. In all cases we observed the development of significant internal know-how. This was always accompanied by a large availability of life-cycle inventory data. The latter can be reached by the development of an internal data-bank (i.e. at BSH, ABB Italy, Perstorp, Akzo Nobel) and/or by the use of sector or corporate data-bases (e.g. ABB Italy using the Swedish data-base and methodologies).

5.4.3.1.3 Companies in the habitualisation stage

Obviously, the companies in this stage (the Swedish company Ericsson and the Italian company Glaxo-Wellcome) are at a very early stage of carrying out LCAs. Therefore, any speculation about their future use of LCA has to be taken with care. However, we think that both cases represent a "good beginning".

At Ericsson, the possible future "success" of LCA is strongly connected with the presence of a very active LCA-entrepreneur and the high organisational learning value of LCA. A couple of LCAs were conducted earlier within various business areas but were not co-ordinated. The entrepreneur had a very clear and effective strategy for creating consensus about LCA within the company. Not only did he initiate LCA activities within his own business area, but also within the corporation as a whole (including also a project to set up a corporate-wide LCA database). He managed to diffuse the idea that LCA is the correct way to look at environmental impacts of products: LCA is product-oriented (which suited him as a product specialist) and is quantitative enough, which appealed to the many engineers at Ericsson.

At Glaxo-Wellcome Italy there are several supporting factors for increasing the use of LCA. The first one is the strong commitment at the corporate level. The second one is the "good" and in some ways surprising results (the LCA demonstrated and quantified that plastics' packaging is more environmentally benign than carton packaging, which was an "environmental fashion" in the early 1990s). Finally, there was a high degree of satisfaction among all people who dealt with LCA.

5.4.3.2 Companies developing alternative methodologies

These are companies which have embraced the concept of Life Cycle Thinking (LCT), but which find LCA too difficult and ambiguous to be used for implementation in product innovation processes. Therefore they are currently developing alternative methodologies for the assessment of the environmental performances of their products. In our opinion, the Swiss company Landis & Gyr and the Italian company Italtel belong to this group.

At Italtel, LCA was firstly tested in 1995, with a study (carried out by an external research institute) on the packaging of a basic telephone model. Despite a generic positive approach to LCA, the conclusions, after the study, were that LCA did not seem appropriate for use within the company's product innovation processes. The main motivation for this was twofold:

LCA turned out to be too complicated to be manageable within the product's decision making processes.

Designers felt that they could influence only in a very limited way the whole product life-cycle (10% only of inventory data came internally from the firm).

The management felt the need for a tool closer to the designer's culture, as well as one complying with emerging market needs. As a consequence, it made a strategic decision, to continue using LCA only for learning purposes and to start developing an internal and alternative tool for environmental product innovation.

As a matter of fact LCA remained at a National Corporate level, where new studies are still being carried out within the Central Engineering department; in particular two more LCAs have been done since 1995 and the Company is internalising more and more specific LCA skills. Italtel's strategy is to continue assessing and testing LCA methodology, while building an internal know how.

On the other hand, a simplified methodology was elaborated, called "Eco-Compatibility Analysis" with the aim of giving a measure of the environmental compatibility of a certain product. In brief, the Eco-compatibility Analysis consists of an evaluation of the employed materials, the project solutions adopted and the fabrication and logistic processes applied to the product. The level of eco-compatibility of a product is assessed thanks to several different indexes then finally summarised in a general one.

The methodology is clearly designed for implementation in daily routines for product innovation: Thanks to the assumptions made for the construction of indexes, the methodology allows product designers to focus on those environmentally negative aspects of products upon which they can have an influence. On the other hand those negative aspects that are not under the direct influence of designers or of the company rate lower. This methodology considers just the product life cycle phases that can be directly controlled and managed by the company. The "Eco-Compatibility Analysis" was immediately applied as a tool for Design for Environment.

Italtel is a clear example of a company which embraces the concept of Life Cycle Thinking, but which needs more simplified tools to translate this approach into implementation routines for product environmental innovation. LCA is still important as a strategic tool and sometimes also for providing data and criteria for the Eco-Compatibility Analysis methodology.

The same situation holds in Landis & Gyr, where no further LCAs are planned, but where an alternative tool for the assessment of environmental properties of products has been developed. The tool, called "Product-Eco-Performance" (PEP) is a sort of checklist with environmental criteria such as efficiency of materials, hazardous substances, packaging, energy use, ability of recycling and life-cycle management. These criteria are rated and added up to an overall ranking note, which assesses the environmental performance of a product. The use by the designer is very simple, as he has to check his design process with this tool and his goal is to reach the minimum possible points of environmental impact. Also in this case, several criteria of PEP are based on information gained by former LCAs carried out at Landis & Gyr.

We think that these two are very important examples. In some way they are extreme cases showing the need for simplified and rapid tools which can be easily used by designers and any engineer. This need is clearly also felt in companies still carrying out full LCAs.

5.4.3.3 Uncertain companies

Two German (AEG and Weleda) and two Swiss companies (KJS and HCB) belong to this group. AEG and KJS are the more interesting cases as they already find themselves at the objectification stage. In contrast, Weleda and HCB are still at the habitualisation stage.

At KJS there is an entrepreneur. People are satisfied and have learned a lot from LCA about the environmental aspects of their products. However a major justification for the further (and systematic) use of LCA internally in the company is missing: "The results of the study suggest that programmes aimed at reducing environmental impacts should, as a priority, address the agricultural stage of coffee production. Actions that could be undertaken by KJS individually respective autonomously proved to be relatively ineffective. As a consequence, KJS took the initiative to set up a co-operation project aimed at the development of best practice in the field of cultivation and processing of coffee beans".

At AEG the situation is different. The early motivation (eco-image of the company) is still valid. Other supporting factors are the continuing pressure of competitors and the potential for using LCA within product development activities. There is an entrepreneur – the Central Technical Department – and he is central enough. However, there are also strong hindering factors, which suggest that the use of LCA within the company may diminish in the future. The first is that there is some internal know-how, but that it is not sufficient to carry out LCA studies without significant external support. The second is that AEG deals with complex products. As a consequence, management currently considers that the costs of carrying out full LCAs are still too high and the benefits too long-term. Moreover, the company recently experienced a significant internal restructuring. Within this process the department formerly responsible for LCA was reformed. Also the formal organisation of product teams, which earlier guaranteed internal communication between different departments was dismantled. In fact, this internal restructuring delayed all activities and has made the future of LCA at AEG rather uncertain.

Weleda is still at the first study stage. Its uncertain position therefore is related to this rather recent involvement in LCA. Important supporting factors for LCA are the original motivation (eco-image) and the presence of strong promoters. Weleda is an SME. This implies a flexible organisation and rather informal and easy internal communication. However this also implies limited resources and with that limited internal know-how. Moreover, one main justification for LCA is missing: On one hand management thinks that LCA has brought new insights in the performance of products, on the other hand they argue that "there is no urgent need for LCA since the products of the company are natural anyway".

5.4.3.4 Companies fading out the use of LCA

The first observation is that in the three cases of Ytong, Cartiera Favini and Ernst Schweizer, no internal know-how and/or organisation have been set-up in any of these companies. The reasons for this might be very different from case to case, however it should be noticed that Cartiera Favini and Ernst Schweizer are small enterprises and that Ytong a medium-size one.

A tentative conclusion seems to be that external consultants may spread the word of LCA, but companies using consultants do not seem to really learn about LCA unless people at the company actively are involved in the LCA study. In short, buying a consultancy report is not an effective way of introducing LCA.

Apart from the lack of sufficient "internalisation" of know-how, there are also other good reasons for stopping the introduction of LCA in these companies. Despite the existence of a good role and justification for LCA (for R,D&D activities), another major and necessary condition was missing recently at Cartiera Favini, that is the mandate and support of top management. In fact, during a recent restructuring, the earlier LCA entrepreneur was decentralised and the priorities in the strategies of the company changed.

The same positive role for R&D activities exists at the German company Ytong (curiously, the R&D was an earlier opponent in this case). However, the earlier motivation of using LCA for marketing purposes to obtain relative competition advantage positions in the branch sector[35] failed. The management realised that LCA results are too complex and cannot be used to optimise the environmental performance of its products (often improving one impact category implies worsening another one). Moreover, the previous entrepreneur (marketing) died recently.

The fact that LCA cannot be used for marketing purposes was one main reason for failure at Schweizer as well. In addition, the prevailing feeling within the company was that LCA results are not clear and cannot be used for decision-making. Finally, the company always relied fully on external know-how (by a consultant and by a student who left the company after the LCA study).

Ciba Speciality Chemicals (CSC) is a case apart. This is a rather unique case in which the use of LCA expected to be stopped after the company reaches the objectification stage. As a matter of fact, the company developed significant internal know-how. There was an early entrepreneur (coinciding with the analyst, who left

---

[35] Incidentally, Ytong is a rather unique example in our case-studies in which a company decided to procede against the decisions of its branch association, the building industry, which had decided not to carry out any LCA.

CSC at the end of 1997). However a major justifying factor was missing. The management concluded that most of the environmental impacts (weighted emissions) of the products were related to energy production processes. This is related to the fact that the synthesis of high-grade pigments necessarily involves relatively high reaction temperatures. The only way to change this would be the quite "radical solution" of producing different kind of chemicals. Moreover, the company perceived that it had very limited influence on the rest of the product life-cycle, both up- and downstream. Therefore, there was insufficient justification to further carry out internally costly full LCA studies. The fading out of the entrepreneur's enthusiasm as well as the fading out of LCA activities as a whole are a natural consequence of these findings.

We think that particularly in this case the *perceived* (and not the real) influence on the rest of the product life chain was a crucial factor for the failure of LCA.

5.4.4 Main factors in the institutionalisation process of LCA

In this subchapter we summarise the results of the case-studies in another way, that is by emphasising the factors for success or failure at any particular stage of institutionalisation.

Table 5.8 summarises the important factors influencing the institutionalisation process according to the empirical data obtained from our case studies.

Table 5.8. Success factors in the institutionalisation process of LCA (based on Tolbert/Zucker 1996)

Pre-institutionalisation stage: Habitualisation	Semi-institutionalisation stage: Objectification	Full-institutionalisation stage: Sedimentation
Internal political arrangement • Top-down approach in I and partly CH (Corporation and/or environmental commitment); • Bottom-up approach in D and S (both are working) • Influence of Corporation in I	Existence of an entrepreneur • Present in all "successful" cases; where missing LCA, has uncertain future • Strong personal environmental commitment	Influence of the entrepreneur • Needed to diffuse the tool in other departments and integrate it into product development • Needed to maintain mandate of top-management
Technical/economic factors • Marketing, particularly in D • Environmental debate ⇒Country specific: high CH, D and S ⇒Sector specific: high in car, chemical, packaging industry • Cost savings	Strategy of the entrepreneur • The strategies can be different, but they include: • Good justification for LCA • Identify organisational problem • Monitoring of competitors • Obtain mandate of top-management	Good organisation for internal communication • Even if routine procedures for LCA use are not fixed, a good communication structure is always present (e.g. inter-department working groups, product-oriented business units)
Policy factors • Waste regulations in CH and D • Public procurement in CH • Very low policy pressures in I and S	Good justification for LCA • justifications can be very different (to innovate or to imitate), but where missing, LCA is going to fail or has uncertain future • openness with respect to unexpected results • demonstrable results needed	Large set of champions • As a consequence of the former point • All technical and product related departments involved
Other factors • Co-operative attitude in CH (companies-research institutes) and in S (Product Ecology Project)	Monitoring of competitors • In all "successful" cases • Important both to enhance relative competitiveness and to imitate competitors	Presence of a formal LCA organisation, development of comprehensive internal know-how, availability of data-banks and software • Present in all cases • Low costs for new "incremental" LCA studies • High level of collaboration between companies and research institutions in S (and at least par-

Pre-institutionalisation stage: Habitualisation	Semi-institutionalisation stage: Objectification	Full-institutionalisation stage: Sedimentation
	<u>Mandate of top-management</u> • needed at this stage, where missing, LCA is going to fail • a long-term environmental commitment is a necessary (but not sufficient) supporting factor; where missing, LCA is going to fail <u>Development of internal LCA skills</u> • Where missing LCA is going to fail or has uncertain future	tially in CH) <u>Other factors (from theory)</u> • Adoption is determined by • Importance and resistance of opponents • Cultural support/ promotion by advocacy groups • Demonstrable results • Low costs <u>Precise role of LCA</u> • Use can be different but is precise
<u>Role of LCA</u> • Mostly for learning • Mostly for confirmation (retrospective use) • With marketing purposes in D and Cartiera Favini, Schweizer	<u>Role of LCA</u> • No marketing • Mostly for internal purposes (R,D&D, bottleneck identification) • Still high learning • Slow shift from retrospective to prospective use	<u>Role of LCA</u> • Use can be different but is precise • No marketing, but other forms of external communication to clients stakeholders and within supply chain management • Mostly for internal purposes (R,D&D, bottleneck identification) • Prospective use • Translation of LCA results into actions is mainly determined by economic factors

Table 5.9. Importance of different factors for the institutionalisation of LCA

Factor / Institutionalisation stage	Habitualisation	Objectification	Sedimentation
1. Internal political arrangements: The importance of creating an organisational structure			
• Top-down mandate with regard to LCA	B	A	B
• Involvement of practitioners	B	A	A
• Existence and centrality of an entrepreneur	B	A	B
• Appropriate internal communication	C	A	A
• Large set of champions	C	B	A
• Development of internal know-how	C	B	A
• Long-term environmental commitment (business strategies considering risks and opportunities)	B	A	A
• Short-term environmental commitment (marketing)	A	C	C
• Promotion of advocacy groups	C	A	A
2. Role of LCA: A good justification is necessary			
• Learning	A	A	C
• Openess to unexpected results	B	A	C
• Precise role of LCA	C	B	A
• Short-term influence on "product" innovation	C	B	A
• Long-term influence on "product" innovation	B	A	A
• Monitoring of competitors	B	A	C
• Application for internal purposes (bottleneck identification, R, D & D)	B	A	A
• Application for external purposes (marketing)[a]	A	C	C
3. Technical factors: Complex, but demonstrable			
• Access to inventory data	C	B	A
• Flexibility of method (adoption)	C	B	A
• Demonstrable results / positive experiences	A	A	A
• Perceived direct or indirect control over the life-cycle (supply chain management)	C	A	A
• Existence of ISO-standards (14040ies)	C	B	A
4. Economic factors: Net benefits have to be expected			
• Low costs of LCA	C	B	A
• Cost saving potentials	A	A	A
• Cost of measures suggested by LCA	C	B	A
• Strengthening relative competitiveness	C	A	A

a) LCA is still valid for supply-chain management and information to customers and stakeholders. Swedish companies use it also as a basis for the environmental product declaration.

Explanation: A = high importance
B = medium importance
C = low importance

Table 5.9 summarises the results in another way. It highlights how the specific relevance of factors changes with time and institutionalisation stage. The factors have been identified based on the case-studies and the instutionalisation theory.

Internal political arrangements:

1. The *mandate of top-management* is an important factor for the institutionalisation of LCAs. The relative importance changes along the process chain. At the beginning it is not too important (we observed that both top-down and bottom-up approaches can be effective). However, it becomes crucial in the objectification stage, where consensus is needed. Once the tool is fully integrated within the company, the mandate is obvious and a direct involvement of top management in daily routines is no longer necessary.
2. The *involvement of practitioners* refers to the spreading out of LCA-activities within companies. At the beginning, no formal rules or staff are required. However, in the next stages of objectification and sedimentation, the relevance of more organised structures increases. Particularly in the sedimentation phase we observed the existence of structured bodies – of course, different organisational types for such involvements exist (e.g. ad-hoc working groups, formalised procedures).
3. The importance of the *entrepreneur* is evident particularly in the mid-stage of objectification. The presence and influence of an entrepreneur is a crucial factor for the success of the institutionalisation process. It's him/her who elaborates the strategy to obtain the consensus within the firm (both the mandate from top-management and the involvement of a larger set of champions).
4. It is clear that without appropriate *internal communication* channels, the message of the entrepreneur can hardly be spread nor can a large set of champions be created. This is not important in the first habitualisation stage, but it becomes crucial in the other two.
5. A large set of *champions*, all through different departments of the company is needed for full sedimentation.
6. The *development of internal know-how* of the method and application of LCA is necessary to support the institutionalisation process. We have observed that no company relying only on external support has continued LCA activities in the long-term. Of course, the importance increases with time and institutionalisation stages.
7. In Table 5.9, we introduced a distinction between a *short-term* and a *long-term environmental commitment*. Whereas the short-term refers to the marketing area, the long-term refers to the business strategies considering risks and opportunities. We found out that companies should have an environmental commitment, otherwise LCA-activities are of a symbolic nature and a short-term fashion. On the other hand, although short commitment might have been a main driver for starting LCA in the past, in none of all studied companies is LCA currently used for marketing.

8. Success for the full institutionalisation of LCA needs support by the *promotion of advocacy groups*. These are people who might not be involved directly in LCA activities, but who support them from outside.

Role of LCA

1. The factors identified within the case-studies stress especially the importance of *learning* from LCAs. A process of learning by doing was identified. Learning is very important in the two first stages; as expected, it becomes less important once the tool is fully integrated in the daily business activities.
2. By opening the perspective of the firm on earlier unknown phases (up- and downstream life-cycle phases of their products), LCAs can lead to unexpected results. In many cases new aspects have to be considered which might entail unexpected results. The *openness to unexpected results* is a factor which determines the company's internal treatment and adoption processes of LCA. It strongly determines the learning process. If people are *not* open to unexpected results the learning process might be stopped and the whole LCA institutionalisation process faded out.
3. One of the major findings of the case-studies is that the *role of LCA* changes as it passes through the institutionalisation stages. We concluded that the role of LCA becomes more and more precise during this process.
4. The application of LCA influences both *short-term* and *long-term product innovations*. This is related to how learning is translated into action. We observed that short-term innovations are determined by economic factors (changes indicated by LCA are adopted only if they provide economic gains as well). Of course, short-term innovations are important in particular in the sedimentation stage. However, companies are much more open to long-term product innovations. For instance, LCA results can influence, also from at the beginning of the institutionalisation process, strategic decisions (i.e. next cycle of investments).
5. The *monitoring of competitors* and their activities influences the application process of LCA. Monitoring is a low-medium driver for starting an LCA, however it might become crucial in the objectification stage, when the entrepreneur has to justify the adoption of LCA to the rest of the company.
6. LCA might be applied for a lot of different purposes. A main distinction is between *internal* and *external application purposes*. Based on the case-studies, we observed a shift from external purposes (marketing) towards more internal applications (R&D,D). This is connected with the current methodological limits of the tool, which basically prevent it from being used for marketing[36]. Of

[36] The actual problem is that comparing a product A with a product B with LCA will mostly lead to ambigous and/or disputable results. All assumptions behind the LCA should accompany the result, which is very difficult to be done in particular for products for final consumers. The situation for industry clients and some other kind of informa-

course, in those cases in which LCA is introduced bottom-up by R&D or environmental departments, the importance of those applications remains constant along all institutionalisation stages.

Technical factors:

1. The importance of the *access to inventory data* becomes more important when a company reaches a higher stage of the institutionalisation process of an LCA. We stress the importance of the availability (and quality) of data. This can either be achieved through the development of internal data-bases (which generally happens in large companies) or by access to sector- or country specific data-banks. We feel that the creation of such data-bases would be a major support factor especially for SMEs.
2. Logical, *positive experiences and demonstrable results* are necessary to justify the LCA-work at any institutionalisation stage.
3. Our case-studies introduced the increasing importance of a *perceived control over the life-cycle* considered. When companies perceive that their influence on the rest of the life-cycle chain is limited, their interest in carrying out long and costly LCAs declines dramatically.
4. The existence of *ISO-standards* for LCA (i.e. the 14040-series) as guidelines is important in the late stages. Their existence support the LCA-approaches of business and the communication among companies who co-operate on this issue.

Economic factors:

1. When reaching higher stages of implementation, the *costs of doing LCA* will become more important.
2. The consideration of *cost-saving potentials* influence the start of LCA-activities considerably. In Germany, Italy and Switzerland, they drive LCAs; the Swedish situation is nearly the same because they are perceived via cost avoidances due to future liabilities[37].
3. The importance of realising competitive advantages and of *strengthening relative competitiveness* is quite an obvious task. Our case-studies showed that the introduction of LCA does not need an economic justification at the very beginning, but in later stages, this demand becomes important. This is justified by the results of the survey because the expected benefits of the use of LCAs are long-term ones. Emerging green markets are also not very important driver for LCAs.

---

tion and education of consumers and stakeholders is different. As a matter of fact, Swedish companies use LCA as a basis for environmental product declarations.

[37] Once more a hint at the higher institutionalisation level of LCA in Sweden.

6 The relationship between business and policy: expectations and implications

In the previous chapters, we have presented results on the application of LCA in business. This experience gained is important in understanding the role which LCA plays within business as an environmental management tool and as a tool for product innovation. Moreover, in addition to business also political institutions, governments and public authorities are aware of the concept and also use it. However, their knowledge of the business practices of LCA and their involvement with it differs in the four countries considered in this report. As shown in chapter 3, they are one important client for LCA-studies.

For instance, the role, which LCA plays for public policy making in Germany is not clear. There are considerably different views on it. The Federation of German Industries [BDI] recently published a position paper on the political role of LCAs, in which its focus is on the question "Intervention or innovation?" and thus refers to LCA (BDI 1998). A concrete example is the German discussion on two packaging LCA-studies[1] carried out recently for the Federal Environmental Agency. Especially the second ongoing study is based on the German Waste Management and Product Recycling Act (KrW-/AbfG) which determines some quotas for packaging. Some months after the start of the second study, in 1997, a debate between the BDI and the Federal Ministry for the Environment started and letters between the Minister and the head of the BDI were exchanged. Information was leaked to the conservative Frankfurter Allgemeine Zeitung, which informed the public on this debate (see FAZ, May 6 and August 20, 1997). The BDI argued that this project was not being carried out according to the ISO 14040. In addition to that, it rejected any governmental decision in the area of environmental policy being based on LCA-studies and their results. However, the Ministry and the UBA argue that the application of LCA can contribute to a more rational environmental policy.

This German example shows that there is a lot of mistrust and fear[2] with regard to the application of LCA by political actors.[3].

---

[1] The first study was published some years ago (see Schmitz 1995), the second is still underway.

[2] In Sweden, the culture in contrast is to talk and discuss with each other and to reach agreements both parties, policymakers and industry, can agree to.

The main focus of the research documented in this book is directed towards the expectations of business and political actors on the application of LCA by business. This aspect has been examined using two different approaches:

- Firstly, we considered business expectations on policy-making activities in the area of LCA (see section 6.1). Are there any relevant relationships between the use of LCA in business and public activities ? If so, what are they? Do policy makers need to act ? From the business perspective, which kind of policy measures/actions should be taken? Do companies (up till now) regard public LCA policy activities as a help or a hindrance?
- Secondly, we looked at the policy level and its views on the business application of LCA (see section 6.2). What are government expectations of the business sector development of LCA, irrespective of government actions? Does government need to act? Does government already know the present business applications and potential benefits of LCA with respect to Ministries particular objectives? Where it knows them, what has it already done?

The results presented in this chapter have been collected by interviews with policy people, by analyses of available policy documents and from the results of our survey (see chapter 4) and of our case-studies (see chapter 5).

In section 6.3, we present some conclusions.

6.1 Business view on policy

The business view on policy is based on the results of the survey and the case-studies. We present them in the following four subsections.

6.1.1 Policy as driver for product innovation and role as stakeholder

Business' product innovations can be stimulated by different drivers. Within the survey, we asked companies for the main drivers for product innovation (see subsection 4.2.5.2). The trend within the four countries considered is quite similar: Product innovations are mainly driven by the market (costs and competition). Legislation pressure is only of medium importance and is ranked at the last position of all drivers in Sweden and at the penultimate position in the other three countries considered.

However, the role of political actors as stakeholders influencing business' activities is strong (see subsection 4.2.1.3). This statement is valid for all four countries and also holds for the present and future influences of policy.

---

[3] Whilst aware of these controversies, they are not reported. We propose considering these in a future project. See also Allen et al. (1997, p. 42ff.), Curran (1997) and Jensen et al. (1997) for a description of the application of LCA by political actors; see section 2.2.

One might conclude that the relationship between policy actions and innovation behaviour is not very clear and at best indirect.

6.1.2 Policy as driver for LCA

LCA-activities can be stimulated by quite different drivers. Within the survey, we asked companies for their motivations to start LCA. Environmental legislation is perceived as a driver of medium importance, also less important than the elaboration of official eco-labelling criteria. Especially in Sweden, environmental legislation is not an important driver for LCA.

However, considering the case-studies, we recognised that the picture is more complex than the results of the survey suggest:

- *Switzerland:* Most Swiss companies of the case-studies indicated that political actions and measures have contributed to carry out LCA. The influencing chain is sometimes longer and more indirect. This was the case of one example (general discussion about sustainable development and the accompanying public environmental activities, concerning the packaging debate). In addition to that, discussions at the ÖBU (Schweizerische Vereinigung für ökologisch bewußte Unternehmensführung), in which the Federal Office of Environment, Forests and Landscape (BUWAL) also participated, were important in pushing forward the development and applications of LCA.

  In three other cases, green public purchasing was more or less a direct driver for starting LCA. Because of public and private procurers' demand for product responsibility and the declaration of products, one company introduced a tool, called PEP, for an environmentally oriented product design partly based on knowledge resulting from LCA. This company also mentioned the development of environmental legislation as a highly influential driver. Although there was no external stakeholder pressure concerning LCA activities from another company, the main and direct drivers for carrying out LCA were the green public purchasing activities in Zurich and their procurement guidelines.
- *Germany:* In most of the examined case-studies, the start of LCA-activities was driven by internal reasons ("eco-image", "environmental awareness"), by public discussions on the relative ecological desirability of chemical substances or by activities of competitors. Only in the case of one company did the packaging discussion and the German packaging ordinance stimulate the application of LCA. However, the German Waste Management and Product Recycling Act (KrW-/AbfG) which introduced in its § 22 a general (and very vague) product responsibility is always considered.

  In the case of another company, whilst public policy did not directly influence the conducting of the LCA, LCA-studies on packaging systems or detergents carried out by the German Federal Environmental Agency helped to make the instrument known to the public and thus to this company.

- *Italy:* So far, the national Italian context has not played role in encouraging companies to start LCA. This situation might change with a new waste law introduced in 1997 and which is applied since October 1998. Italian companies, in particular the ones analysed in our case-studies, referred directly to the EU political context or to general environmental concerns or international agreements (e.g. the Rio Conference) when starting their LCA strategy. In particular, the two most mentioned acts are the EU Packaging and Packaging Waste Directive (94/62/EEC) and the European Ecolabel (Regulation 880/92) and the debate on priority waste streams in the early 1990s[4].
- *Sweden:* The interviewed Swedish companies never mentioned LCA in relation to legal frameworks. They work with LCA because it is important for business development and the market communication of the company. When asked about their view on the future introduction of LCA-related regulations, they think it would be a shame. It would probably lead to LCA activity becoming "just another regulatory requirement" and the company losing its own drive to carry out LCAs. The Swedish companies do not want the government to make it mandatory to do LCAs. The companies have realised that they all do their own kind of LCAs in accordance with their own personal goal of study.

The general conclusion is that it is especially the expected or realised product responsibility that stimulates the application of LCA. Public purchasing in the sense of environmental conscious demand could also push companies to check the environmental burdens of their products. The concrete influence however depends on the market power of demand and the competitive situation. An influence from policy on business to conduct LCAs seems to exist in Switzerland and – less so – in Germany. Italian and Swedish policies do not seem to have stimulated the LCA-activities of business.

Considering the reverse chain management model developed by Smith et al. (1998), most of the measures which have stimulated business to carry out LCAs were measures for the improved management of waste and resources. However, the stimulus given by ecolabels goes ahead because it expands the view towards consumers.

6.1.3 Public policy's LCA activities as a help or hindrance?

There are many areas in which policy can act in the area of LCA. However, before looking at the desired policy actions, the question arises as to whether business wishes such activities or not. Are such activities regarded as a help or a hindrance for business LCA-activities?

The situation in the four countries seems to be similar:

---

[4] This debate was introduced by the Commission in 1990. Its objective is to cooperate between the Commission and interest groups that "directly influence production and consumption patterns which generate waste or emissions, for environmentally friendly solutions as far ahead as possible of the waste stage" (Erasmus 1993, p. 5).

- In *Switzerland*, no hindrances can be assessed from public policy's LCA activities referring to business LCA applications. However, public policy's LCA activities were not explicitly mentioned as a help or supporting driver for the conducted LCAs. In spite of this, public authorities have several times been involved in carrying out LCAs and have published data inventories in some areas (see section 6.2). According to future applications of LCA, public policy's LCA activities are seen more as a help than a hindrance; especially when focusing on simplifying the method and the data-base construction. In order to make progress in the evaluation of inventory results, public authorities' efforts in this direction are highly appreciated.
- In *Germany*, the view on public policy's LCA-activities is unclear at the moment. Most of the companies regard them as neither a help nor a hindrance. However, there are a lot of more ideological, discussions between the government, the Federal Environmental Agency and industry on the role of LCA for environmental policy. Most parts of industry emphasise the principle of co-operation in the field of LCA.

  However, the BDI, the Federation of German Industries, does not want policy to take decisions on the basis of LCA-results (see BDI 1996, p. 2f.)[5] The BDI favours the application of LCA within business decisions which encompass for example strategy development, product development and marketing. LCAs are not considered to be an appropriate governmental control instrument and public policy is expected to initiate LCAs in order to bring together the different interested circles[6]. In its recent published position paper, this strong refusal of the BDI has been reformulated once more: "Beyond their role in enterprise, LCAs can also provide orientating knowledge for policy-makers active in the environmental field. They do so by expanding the environmental map with parts of the ecological profiles of products" (BDI 1998, p. 4). "(...) LCAs do not provide a suitable basis for state intervention in the market, for example, with respect to applying taxes, quotas, limiting the use of a product or other instruments of environmental regulation. Static legal intervention based on simplified conclusions can only erode the opportunity to improve product systems dynamically using the sophisticated information obtained by LCAs" (BDI 1998, p. 5)[7].
- In *Italy*, these activities were not explicitly mentioned as a help or a hindrance. The reason might be that the Italian public sector has not played any role since LCA has been applied by Italian companies. Only recently, some awareness has arisen (see section 6.2).

---

5 See for example the debate on the LCA-study on packaging II (packaging systems for mineral water, juices, wine).

6 See Becker/Marsmann (1998).

7 According to our opinion, this (excerpt) of the statement of the BDI is inconsistent: The BDI accepts the application of the results of LCA for environmental mapping in policy. But after this, the BDI rejects any consequences for environmental policy.

- In *Sweden,* the situation is very similar to Switzerland; no hindrances can be assessed from public policy's LCA activities referring to business LCA applications. Public policy's LCA activities were not explicitly mentioned as a help or supporting driver for the conducted LCAs, either. In spite of this, public authorities have been involved several times in carrying out LCAs and have published data inventories in some areas (see section 6.2). According to the future applications of LCA, public policy's LCA activities are seen more as a help than a hindrance.

One might conclude that there are different policy "styles": The Swiss one characterised by harmony and accepted support, the German one characterised by conflict and debated support, the Italian one characterised by modest awareness and very limited support[8] and the Swedish one characterised by harmony and accepted support.

6.1.4 Desired policy actions

The desired support of business by policy and public institutions exists in different ways in the four countries studied (see subsections 6.2.5 and 6.2.6). But which desires and expectations do companies have? Which measures are claimed?

Clearly, the expectations depend on the national context, the national policy "style" and national public activities. However, beside the more fundamental question of help or hindrance, a series of different possibilities to support the application of LCA within business exist. Therefore, we have also considered these aspects within our 20 case-studies[9]. The results are:

- *Switzerland:* Two case-study companies think that support of research on methodological simplifications and on standardised interpretation/evaluation methods is necessary. They also ask for further development of data-bases and for standards for general material codes and product data codes. Also financial incentives for environmental behaviour are welcome. It was interesting to notice however that three other case-study companies did not ask for any public support at all.
- *Germany:* Most of the five companies which we considered ask for the support of research in the area of impact assessment and interpretation and especially for the creation of public data-bases. Other mentioned desired actions are: the establishment of a general product classification with classes according to the environmental-friendliness of products[10], support in the further standardisation

[8] However, the situation is likely to change because an Italian Association for LCA has been founded by the beginning of 1999. Its members are researchers, employees of different Italian companies, and representatives of the Ministry of the Environment and of the Environmental Agency (ANPA) and NGOs.

[9] Beside that, some interviews have been carried out to get more information.

[10] This general classification could work in a similar way to the energy-efficiency classes of electrical household appliances, but by taking all environmental impacts and the entire

especially of goal definition and interpretation, aupport in a detailed standardisation of impact assessment and interpretation by prescribing impact categories that have to be considered in every LCA-study, providing fixed weightings for the different environmental impacts. It was also mentioned that further participation of public authorities in standardisation activities is necessary. The application of LCA should be supported by promoting LCA especially with regard to SMEs, by giving subsidies for LCA-application and by providing information and know-how for companies. Surprisingly, the BDI (1998) does not claim any wish for concrete support or policy action with regard to supporting business' LCA-activities.

- *Italy:* The Italian companies considered focus most of their expectations nowadays on the Italian LCA association and regard this as an effective possibility to solve current problems. Some of them asked for public support in creating an official and transparent data-base, in including environmental requirements in public calls for tenders and in supporting official eco-labels. ANPA is presently tackling with this issues, as it commissioned an important LCA study which should lead to the establishment of a public National data-bank on Energy, Transportation, Waste Management systems and main building materials. The first version is to be ready at the beginning of 1999. The official final version should be finished later this year or at the beginning of 2000.
- *Sweden:* The Swedish companies considered did not make any specific demands with regard to public activities and support. However, additional interviews showed that further improvement of data-bases and of data-base formats would be welcome. Other expectations are the development of environmental product declarations and of the ecolabel according to ISO-type III[11] and of the improved teaching of LCA at universities to increase the number of educated personnel available.

Before drawing general conlusions we have to remember to it restricted empirical basis. Nevertheless, it is clear that the development and improvement of databases is claimed in all countries as a very important measure. Swiss and German companies wish ongoing methodological support, especially with regard to the weighting and interpretation of the results of the inventory.

---

life-cycle into account. The expert of the company who made this proposal regards LCA as the appropriate instrument to provide the information required and expects public policy and particularly the Federal Environmental Agency to take an active part in this context.

[11] The recent proposals for the ISO-type III ecolabel describe it as follows: "A Type III Environmental Declaration is a quantified environmental life-cycle (cradle to grave; or cradle to gate) product information profile, provided by a (supplier), based on life cycle assessment (LCA) according to the ISO 14040 series critically reviewed by a qualified independent third party, presented as present categories of parameters for a sector group" (ISO/TC207/SC3/WG1/TG Type III N 47, p. 1). The standardisation development is still under progress.

In Germany, it is also expected that public authorities will also join in future ISO-standardisation activities. This might be interpreted as a hint for consensus-building activities. Also important is the support of the application of LCA by small and medium-sized companies.

The application of LCA by business does not happen in a policy-free area. Companies are aware of the political framework and the interviewed case-study companies ask for product policy actions which strengthens their environmental approaches. Some of the expectations are concrete and based on existing tools (support of official eco-labels, environmental orientation of public call for tenders). Some measures are more long-term (especially the proposal by one German company to establish a general product classification made according to the environment-friendliness of products). One has to doubt if these claims are in the "mainstream" of business' expectations. But they represent also some actual interests.

6.2 Policy view on business

6.2.1 Introduction

The life cycle concept is a powerful system approach for thinking about technology and products from a "cradle to grave" perspective. A more widespread application of the life-cycle concept in business may improve the public policy process by providing more information to decision-makers in a comprehensive manner. In fact, environmentally effective public policy addresses the entire range of potential environmental impacts along the chain of commerce associated with the use of products, processes or activities.

6.2.2 The European perspective

The pressure to assess the environmental impact of products and processes has been more severe in Europe than elsewhere, especially in northern European countries. Policy makers started to refer to life-cycle concepts or cradle to grave approaches already at the beginning of the 1980s (e.g. BUS [1984)]. The activities of European regulator and regulatory agencies have also been a major factor influencing both in the pace and the nature of LCA development in Europe.

The most common application of Life Cycle Thinking is in the development of product oriented environmental policy, also called „Integrated Product Policy" (IPP); an important area are eco-labelling programs. The Council regulation 880/92, for example, requires that specific ecological criteria for each product be established using a 'cradle to grave' approach.

The first example of a European regulation with a direct and explicit reference to the term 'LCA' is the Packaging and Packaging Waste Directive (94/62/EC). Its preamble states: "(...) life-cycle assessment should be completed as soon as possible to justify a clear hierarchy between reusable, recyclable and recoverable packaging". In this way the Packaging Waste Directive sets targets for recycling and the recovery of packaging allowing these to be changed according to the results of an LCA.

Many other examples are present at the national level, with most of them related to waste management. In The Netherlands, for example, the Packaging covenant requires: "(...) If research shows that replacement of one-way packaging by re-useable packaging would clearly cause less damage to the environment and there are no preponderate objections to such a change-over (...) then packaging industry will undertake to switch over to using re-useable packaging rather than one-way packages". After three years of the implementation of this covenant, LCAs were produced for some ten different product categories. Meanwhile, industry developed a number of novel packaging options with a view to the reduction of packaging waste output. LCAs helped industry to recognise improvement opportunities, rather than choose between options existing at the time of signing the covenant.

Encouraging innovation of greener products has become the main focus of Dutch integrated product policy. With the proposal for developing "product oriented environmental management systems" (POEM) the Dutch policy aims to, amongst other things, put in place an integrated chain analysis involving the complete product life-cycle (e.g. LCA).

The Danish EPA is promoting the development and transfer into industry of the Environmental Design of Industrial Products (EDIP), an LCA based evaluation tool. Similar activities have also been sponsored in Germany, Austria and Finland.

The attention of R&D efforts towards a widespread diffusion of LCA practices is also recognised in the second amendment proposal concerning the "5th Framework Programme of the European Community for research, technological development and demonstration activities" (EC 1998a) for the years 1998 to 2002. In fact, amongst other priorities, research will focus on: "(...) technologies to reduce resources utilisation and to promote the reuse and recycling of waste and the development of clean processes and products based on the concept of 'life cycle analysis' (...)". The terms "Life Cycle Approaches" or "Life-Cycle Impacts" are also mentioned in other draft documents concerning the specific programmes implementing the 5th Framework Programme. In particular, the terms have been recalled when discussing the sustainable use of raw materials, the transport of goods and passengers and eco-efficiency processes[12].

A major effort has been undertaken by the Directorate General for Environment, Nuclear Safety and Civil Protection (DGXI) towards what they call "Integrated Product Policy (IPP)". IPP is a relative new area in environmental

[12] See for example EC (1998b).

policy, which is receiving increasing attention in several EU Member States as well as in international organisations like the OECD and the UN Commission on Sustainable Development. The global objective of IPP is to improve resource efficiency and reduce environmental impact across more than a single stage of the product life cycle. Ideally it would be concerned with the total life cycle. A recent report[13] on IPP on behalf of the Directorate General XI of the European Commission recommends that emphasis should be placed on R&D, supporting 'green' product development. This could be done by means of:

- supporting the development of data, analytical tools and methodologies for applying life cycle environmental knowledge to the development of new products; and
- facilitating the adoption of Life Cycle Approaches and product management systems by European firms.

Recently, an Informal Council of the European Ministers for the Environment confirmed the issue of IPP[14].

6.2.3 General and LCA related policy principles

The general principles of environmental policy in the four surveyed countries go along with the model of sustainability[15]. There is a clear shift from traditional command and control policies to market-based instruments and voluntary agreements in the form of co-operation between industries and public authorities. 'Polluter Pays Principle' (PPP) together with 'producer responsibility' are the main principles of the new environmental policies of the 1990s.

However, the applications of these common principles at the European level are taking different forms and patterns of implementation according to the country considered. Northern European countries have been keener on a more systematic and comprehensive application of these principles (including the 'precautionary principle'). Southern countries tend towards a soft application of the same (i.e. shared responsibility[16] instead of extended responsibility and soft environmental charging systems).

The principles which refer to the business' application of LCAs are both the introduction of Life Cycle Thinking in companies and the help-for-self-help and good-example-strategy in order to achieve sustainable development in production, products and finally consumption. Within the spirit of sustainability all efforts in this field aim to close substance loops in the long run[17].

---

[13] Ernst&Young/SPRU et al. (1998, p. 77).

[14] See BMU (1999).

[15] See Knoepfel (1997, p.190), UBA (1997b, p. 177), Ministry of Environment (1997).

[16] See for example the Italian Waste Act (N.11/97), Article 41, c.2h

[17] The German Waste Management and Recycling Act (KrW-/AbfG) has been created to support these goals.

6.2.4 Public LCA activities

Public LCA activities are mainly focused on supporting or revising environmental regulations at European and national levels. As a matter of fact, most of the national and European public agencies are increasingly using Life Cycle Thinking (and some of them specific LCA) as a tool for developing environmental policies strategies.

The use of LCA for public policy has led to wide controversy. At the European level, for example, the LCA (RDC / Coppers&Lybrand 1997) conducted as a part of the follow-up of the Packaging Waste Directive created a lot of discussion and confrontation between policy makers and industry. The need for the public peer review of any LCA intended to be used for policy making has been stressed at the national level both in this and other similar cases (e.g. the packaging study known as 'packaging II' financed by the German Federal Environmental Agency [UBA]). As a result of these controversies, the UBA committed itself to carrying-out LCA-studies (UBA 1998); this commitment is addressed especially towards business and industry.

The main activities by the countries considered are summarised below:

- *Switzerland:* The Swiss Agency for the Environment, Forests and Landscape (BUWAL) plays an important role in developing and supporting LCA-activities. Amongst the activities of the agencies we can mention the support for the development of data-bases and the involvement in carrying out several LCAs on transport systems, building materials, agricultural products and packaging.
- *Germany:* The Ministry for the Environment (BMU) and its scientific consultancy, the Federal Environmental Agency (UBA), are very active in the area of LCA. The Ministry regards LCA as a tool for environmental policy (see Schafhausen 1998; Troge 1997). Exemplary questions which might be answered by support of LCAs are:
  - Which type of waste treatment has environmental priority?
  - Should environmental policy support electric cars?
  - Should regenerative energy sources be subsidised?
  - What are the conditions (e.g. return quota, distribution distances) for environmentally preferable packaging?

These questions have to be seen in the context of environmental legislation and policy, especially eco-labelling, the German Packaging Ordinance with the rules of return and collection quotas, and the legislation for extended producer responsibility within the Waste Management and Product Recycling Act (KrW-/AbfG). That means that the results of publicly financed LCA-studies are used for setting up the institutional framework of product policy measures for the business world (e.g. return and collection quotas, eco-labels, recommendations for public procurement and financial incentives).

Therefore, UBA financed a lot of LCA-studies of which the first objective was to use the results for policy-making or consumer information; only a minor objective was to support business by offering inventory data. Examples (see UBA 1997a) are:

- LCA on bags made from polyethylene, paper and jute (UBA 1988)[18];
- LCA/product line analysis on detergents (Grießhammer 1997);
- LCA on rape oil (UBA 1993)[19];
- LCA for drink packaging systems ("packaging I" and "packaging II"-studies)[20];
- LCA in the area of waste treatment, e.g. waste tyres, refrigerators and waste paper;
- LCA for eco-labelling activities (e.g. detergents).

It is important to mention, that in Germany, based on some years of research and consulting, the draft of a unique environmental basic law has been published (BMU 1998b). This proposal has been commissioned by the BMU and intends to combine all the different environmental laws, ordinances, prescriptions etc. within one single law. § 119 refers to an "Environmental burden analysis"[21]; this paragraph proposes that government is allowed to force producers of specific products to carry out an environmental burden analysis. Other very general prescriptions (§§ 118 and 123) claim an environmental product responsibility. However, experts doubt if these prescriptions will really come into force.

- *Italy:* Public authorities have only recently begun some activities in relation to the use and support of LCA. These are mainly connected with the eco-labelling programme and the development of national data banks. In fact, the main on-going project in Italy is the development of a national data-base for LCA commissioned by the National Agency for the Protection of the Environment (ANPA). This project consists of data research, analysis, aggregation and adaptation both at the European and national level. It particularly focuses on on Energy, Transportation, Waste Management systems and main building materials. The first version is to be ready at the beginning of 1999. The official final version should be finished later this year or at the beginning of 2000. Moreo-

[18] This study which was not really an LCA was discussed very heatedly because NGOs have been in favour of paper and jute-products which ranked worse than plastic bags.

[19] This study came up with the environmental result that rape oil is better than mineral oil in most of the impact categories. But then the UBA introduced an economic criterion, namely the relationship between costs and benefits and concluded that an intensification of rape oil would not be the best investment for reducing CO_2-emissions and advised the Ministry not to do this (anymore). Of course, agriculture was not happy with this recommendation and contacted the Ministry for Agriculture.

[20] The packaging II-study is based on the background of the German Waste Management and Product Recycling Act (KrW-/AbfG) which determines some quotas for packaging. The first stimulus for this study was a critique of business at the packaging I-study. Therefore, an improved update was agreed on.

[21] The "Environmental burden analysis" is a new term invented by the authors. They seem to intentionally avoid the term "LCA", but there are no differences between the terms.

ver, other specific examples of LCA carried out and/or being requested by the ANPA are LCAs for refrigerators, ceramic tiles and packaging (in connection with eco-labeling activities). The agency is also planning to start LCA related activities on waste management options[22], in particular on plastic recycling in collaboration with the University of Palermo.

Some new activities in the area of methodological development are also under consideration such as on streamlined LCA and Life Cycle Costing (LCC).

- *Sweden:* Public authorities in Sweden have played a very active role in the development and use of LCA for building policy strategies and recommendations. The Ecocycle Commission[23], for example, made an extensive analysis to assess the environmental impact of the use of PVC. The method used was an LCA. Based on the results of this analysis the Commission recommended the phasing out of this material. During the period from 1989 to 1992 a special committee on packaging used an LCA to suggest packaging targets and producer responsibility.

  However, we could not track down any recent examples of LCAs carried out for policy making activities, one reason seems to be the high cost for performing complete LCAs (i.e. including the assessment stage). According to the Swedish Environmental Protection Agency the application of LCAs in the near future will be limited mostly to the inventory stage.

It seems that the primary objective of public LCA activities has been to use their results for policy making or consumer information; the support of business by offering inventory data remained only a minor objective (although this is a main goal of the project of ANPA with respect to the establishment of a National databank on energy, transportation, waste management systems and building materials in Italy).

In contrast, the policy development in Switzerland and Sweden shows that a quite active role of policy makers is to support the business application of LCA in companies (and other actors).

6.2.4.1 Support of scientific work in universities

In most of the surveyed countries public national authorities are supporting activities by means of financing programmes for PhD students or supporting, through research programmes, the development and refinement of LCA in the context of weighting, evaluation and interpretation. The overall goal is to contribute to the methodological progress of LCA.

In Switzerland, for example, within a special research programme called 'SPP-Umwelt', several LCA-projects have been financed. The Environmental Protection Agency (UBA) in Germany, as well, is supporting projects on LCA method-

[22] ANPA, personal communication to IPTS, 24 August 1998.

[23] This was set up by the Swedish Government in 1993 after the approval of the Ecocycle Bill.

ology. The Technical Environmental Planning department (TMP) at Chalmers University is the only department in Sweden with its main focus on LCA. The unique position of TMP has transformed it into a real reference centre, with established contacts with most of the doctoral students in Sweden.

6.2.4.2 Support of agreements on weighting and valuation methods

In Germany, in contrast to other countries, especially Sweden and Switzerland, there is a broad consensus among different interest groups that methods which summarise environmental burdens into an overall score are misleading. Therefore, the UBA sponsored two reports in the area of impact assessment and interpretation (Giegrich 1995 and Kloepffer et al. 1995).

These reports contributed to the organisation of a workshop on the valuation within an LCA which took place in June 1996. Following this workshop a project started in which the original intention of the client (UBA and Ministry for the Environment) was to arrive at a consensus on the valuation/interpretation/weighting-step among different interest groups and stakeholders who joined the project as a working group. This project failed its objective. A new definition was agreed upon. The outcome was a set of agreed and consensual principles within an LCA which do not restrict the methodological pluralism and describe more the process of carrying out – if necessary – an interpretation within an LCA.

Another way of supporting business is the development of environmental quality and action goals which signal environmental priorities to the public and business.

6.2.4.3 Support of development and publication of data-bases

This is of the most active area of development in countries like Italy, Switzerland and Sweden. Germany has also done some work along these lines, however most of the LCA studies on behalf of the UBA (e.g. the development an LCA module for energy systems and publishing LCA-studies via Internet) have not had the primary intention of delivering inventory information to business. The primary objective has in fact been to support environmental policy in general.

As mentioned above, the Italian Agency for Environmental Protection (ANPA) has recently founded a project for building the first Italian data-base for LCA on energy, transportation, waste management systems and building materials. In Switzerland, BUWAL has supported the development of data-bases which can be used by everyone for LCA-related purposes (e.g. packaging, energy systems and energy sources, transport systems, etc.). A new WEB page supported by BUWAL contains a list of LCAs carried out in Switzerland during the last years and a list of publications[24].

[24] URL: http://www.admin.ch/buwal/projekte/biotech/lca/

In Sweden, the Centre for Environmental Assessment of Product and Material Systems (CPM) made a major effort to establish an LCA data-base with a standardised LCA format (SPINE). This format is used by the CPM founder companies both in their internal data-bases and in the data-base at CPM where the companies share LCA data with each other. The project is founded both by public (Swedish National Board for Industrial and Technical Development – NUTEK) and private institutions (e.g. Volvo, ABB, Electrolux, etc.).

6.2.4.4 Support of environmental management systems

The introduction of environmental management systems according to ISO 14001 and the European EMAS scheme have been supported in countries like Germany, Switzerland, Sweden and more recently in Italy. This support has been granted on the basis of the positive expected impact of implementing environmental management systems (EMS) in companies.

Policy makers (OECD 1998b, p. 17) believe that an effective EMS should enable companies to:

- anticipate and meet growing environmental performance expectations;
- ensure ongoing compliance with national and/or international environmental requirements; and
- support the continual improvement of their environmental performance.

6.2.4.5 Financial support

In all countries surveyed specific national programmes which offer direct financial support to companies for LCA activities do not exist. In particular, no national support is present, to our knowledge, to SMEs willing to start-up LCA activities.

6.2.4.6 Personal support

Some of the employees of UBA (Germany) and BUWAL (Switzerland) are very active in the area of LCAs. They participate in meetings and conferences and support business through internal and external working groups, review panels or even by giving direct assistance.

6.2.5 Expectations of policy makers towards business application of LCA

The general expectation of policy makers (at the European and national level) towards business application of LCA is that companies are now increasingly aware of environmental burdens along the complete life-cycle of their products.

The use of an analytical tool such as an LCA is seen positively as an opportunity for introducing LCT in the business perspective. Other and more concrete expectations of policy on the business application of LCA are the:

- potential of the instrument as a tool for developing indicators that may used by industry for measuring their performance[25];
- use of LCT for supporting 'Eco-efficiency' concepts and application[25];
- potential application for strategic decisions (i.e. new products' decisions based on comparison between different products);
- identification of improvement potentials of products (based on analyses of single products);
- raising of environmental awareness in companies;
- identification of environmental priorities[26];
- support of the introduction of environmental objectives in companies[26].

Policy makers have generally supported LCT in business however some important requirements have been stressed to gain a widespread diffusion of LCA practices, as for example:

- transparency and verification;
- adapting LCA to customer needs and the complexity of the product or process considered;
- simplifying the approach as much as possible, more towards 'check list' than complete LCA application;
- flexibility and harmonisation of methodologies (comparability of results for using LCA in public policy making).

6.2.6 Other activities

6.2.6.1 Public administration internal LCA-groups

It has been reported that both BUWAL in Switzerland and UBA in Germany have set up internal ad-hoc groups on LCA. In Italy, an association for the development of LCA methodology and application, has been officially formed very recently. The association includes representatives not only from the public sector but also from major Italian companies[27] as well as representatives from the academic world.

[25] European Commission, DGIII and DGXI, personal communication to IPTS, 18 June 1998.

[26] Swedish EPA, personal communication to IPTS, 24 August 1998.

[27] Three of our case-study companies are participating in this, namely Fiat Auto, Italtel and ABB Italy.

6.2.6.2 *Public green purchasing based on LCA-information*

In Switzerland public authorities have a quite strong commitment to purchasing products according to environmental criteria such as office material (paper, pens, furniture), building materials (paints, concrete, insulation), cleaning agents for buildings, goods for hospitals etc. In some areas LCA has been used for the evaluation of criteria for environmental purchase. (e.g. paints and varnishes, sealing of parquets). This procedure of integrating LCA-based considerations in the requirements for suppliers might stimulate the application of LCA in companies because they have to demonstrate if and how they fulfil the demanded standards.

In Germany, green public procurement is not mandated, but is stimulated by government. In particular, information is provided in the form of a "Manual for Environmental Procurement". This manual, which gives decision support in the area of environmentally responsible procurement, is available to Ministries and interested companies. At the start of this year, the government published the forth edition. The methodology used to make product comparison is, where possible, LCA.

In Sweden, the Public Procurement Act provides new opportunities for green purchasing, however the scope and content of requirements specified in the act are as yet unclear and may be linked to eco-labels. A public procurement board has been set up to monitor all public procurement and to provide guidance and advice. Many municipalities and regional governments have independently developed handbooks and practices.

6.2.6.3 *Standardisation*

An employee of UBA heads the working group "LCA" of DIN. The whole standardisation work in the area of environmental management, eco-labelling and LCA is subsidised by the BMU.

It is also intended (see BMU 1998a, p. 24) to agree on a code of conduct for LCA which contains some arrangements for methods and procedures in the framework of LCA. However, it is not clear at which level and between which organisations this code of conduct should be agreed upon.

6.3 Conclusions

The role and the perception of policy makers on the use of LCA in business decision-making, as depicted in the two previous sections 6.1 and 6.2, have taken different patterns and directions in the four countries considered. Undoubtedly European and national governments generally support activities and actions directed towards the minimisation of the impact over the whole life-cycle of a product or process, but not necessarily by using a tool such as LCA.

The policy patterns with regard to the application of LCA within public policy domains seem to differ in the four countries. In Switzerland and Italy, LCA is not or rarely used within public policy making. The German situation in contrast: LCA is applied by the Ministry and the Federal Environmental Agency within policy-making (especially in the waste area) and a lot of discussions and conflicts between government and business/industry have been the consequence. The Swedish situation is influenced from the culture of talking and discussing with each other; the application of LCA within the Ecocycle Commission seems to be less controversial and more harmonious.

Altogether, the main areas of the application of LCA/LCT within public environmental politics are:

- Waste treatment options;
- means of transport;
- energy sources; and
- products' choice.

Policy has very general principles referring to the business' application of LCA; these are both the introduction of 'Life Cycle Thinking' (LCT) in companies and the help-for-self-help and good-example-strategy. The latter aims to achieve sustainable development in production, products and finally consumption. Life Cycle Approaches and/or LCT are regarded as a key factor in the future development of green product innovation and a key element in the a widespread take-over of sustainable development both in public and private institutions. The use of an analytical tool such as LCA is positively seen as an opportunity to introduce LCT in the business perspective. Policy's expectations of the business application of LCA are:

- the potential of the instrument as a tool for developing indicators that may be used by industry for measuring its performance;
- using LCT for supporting 'Eco-efficiency' concepts and application;
- the potential application for strategic decisions (i.e. new products' decisions based on the comparison between different products);
- identifying improvements' potentials (based on analyses of single products);
- raising environmental awareness in companies;
- identifying environmental priorities;
- supporting the introduction of environmental objectives in companies.

However, governments do not seem to have a clear picture of the real business' application of LCA yet; therefore the measures which need to be taken to support business are still not co-ordinated and structured enough into formal programmes. Nevertheless, there are a plethora of different measures and actions which have been taken/considered by governments in the four countries. They have been involved either directly, financing LCAs to be used for their policy processes or indirectly, supporting the implementation of LCA in industry as an instrument for achieving sustainable development and financing projects for methodological refinements. Worth mentioning is the absence of support in the use of LCAs in

SMEs. It seems that they are increasing look for assistance in integrating LCAs in their business management to answer both market and legislative pressures.

The application of LCAs within business is partly stimulated by policy, especially by expected or realised product responsibility. Public purchasing in the form of environmentally conscious demand might also push companies to check the environmental burdens of their products. This concrete influence depends on the market power of demand and the competitive situation. The influence of policy on business to conduct LCAs seems to exist in Switzerland and – less so – in Germany, whilst Italian and Swedish policies do not seem to have stimulated LCA-activities of business.

The expectations of business towards policy support are embedded in the national policy "styles"; we identified three variations:

- harmony and accepted support (Switzerland and Sweden),
- moderate awareness and limited support (Italy) and
- conflict and debated support (Germany).

Business' expectations in the countries considered present one main (and not surprising) consensus: the support of the building up of official and public databases. Also methodological support in the area of impact assessment and interpretation is desired in three countries. Some of these expectations have already been recognised by policy makers who have identified some important bottlenecks that prevent the widespread diffusion of LCA in industry as for example the:

- need for transparency and the possibility of verifying LCA results;
- rigidity of the tool LCA and the standard ISO 14040;
- difficulty of adapting the LCA tool for the analysis of complex products and;
- missing or insufficient public available data banks which support business with environmental product related data.

In dealing with the latter, major research projects have been sponsored by public institutions in Switzerland, Italy and Sweden. Modest in comparison are the efforts by the German public institutions in the field of public data-banks.

7 Conclusions and recommendations

In this chapter 7, we present our general findings and main conclusions. The results are summarises in section 7.1. Section 7.2 summarises the main findings. Recommendations are presented in section 7.3.

7.1 Summarising results

7.1.1 Introduction

The history of the tool Life Cycle Assessment (LCA) is a relatively new one: beginning in the late 1960s, its real breakthrough into the business world occurred in this decade. Enthusiasm and scepticism alternate when speaking about LCA: LCAs have been hailed as the "environmental management tool of the 1990s", "LCA will be seen as an integral part of the environmental management tool-kit" (Jensen et al. 1997, p. 28). Arnold (1993) and White/Shapiro (1993) think that LCA procedures are too expensive and complicated; hence they can only be seldom used. Ehrenfeld (1997) discusses several pros and cons. Contrary to this, Berkhout (1996, p. 151) mentions: "The adoption of life cycle approaches is a sign of beyond compliance behaviour". A lot of controversy in the scientific discussion exists. Nevertheless, the progress on the ISO-level is impressive: two standa2512rds have been agreed upon, several other are underway[1].

LCA is a tool for assessing the potential environmental impact of products[2]. It assesses products from several perspectives, taking into account:

- the whole life cycle of the product, i.e. combining an upstream with a downstream focus,
- consumption of different input streams, e.g. non-renewable resources,
- cross-media outputs, i.e. considering environmental burdens within several ecological media.

Its applicants are a plethora of different actors and groups, such as business, business associations, government, public agencies, NGOs and researchers. Who are

---

[1] See Box C for an overview on the actual state of ISO-14040-series.

[2] As mentioned above, it has to be remembered that the term "product" should be interpreted in a broad sense and includes also services.

the most important applicants? The answer is clear and unambiguous: in chapter 3, we showed that business[3] predominates the commissioning bodies in all of the four countries we examined; their share makes up more than 60% in Germany, Italy and Sweden[4].

7.1.2 The institutionalisation of the tool LCA – Who uses LCA – why and how – within business decision-making processes?

Institutional scripts and rational choice

The ways and paths by which LCA is introduced in companies can be very different. The process of adopting LCA within a company's objectives is influenced by different factors, and these can determine the whole process of either success or failure. This process of introducing and adopting a new tool (and also, more important, new views) into a company is called "institutionalisation" and has been analysed within the framework of organisation theory. Tolbert/Zucker (1996) present an overview of this institutionalisation theory. They describe the process of adoption to new situations, requirements etc. as the pursuit of an institutional "script"; hereby adding new insights into adoption processes within companies beside the model of rational choice. The institutionalisation theory describes the characteristics of different phases in the introduction of a new phenomenon until it becomes something taken for granted. It also describes factors important for a complete institutionalisation. We have used and "translated" this theory to develop a framework to analyse the process of the introduction of LCA into companies. The institutionalisation theory distinguishes four stages:

1. Innovation,
2. habitualisation,
3. objectification and
4. sedimentation.

This theoretical framework was applied by us to the process of analysing the adoption of the new "phenomenon" of LCA.

Which are the triggers and drivers? From innovation to habitualisation

Institutionalisation theory explains that LCA-activities only start in companies if an organisational problem is recognised. The problem might be an inappropriate structure for assessing products or environmental demands which cause new challenges for companies.

[3] Taking into account the requests for confidentiality by business as LCA-clients, we think that its importance as a client and commissioner of LCA-studies is considerably higher.

[4] Only in Switzerland, with 42%, was the share of business as commissioners more modest.

According to the institutionalisation theory, the adoption of LCA-activities within companies is determined by economic, technical and company-internal political arrangements. This is verified by our case-studies. Some of our case-study companies started LCA-activities because they intended to use the results in marketing. Another economic factor at this stage was the search for cost savings. Another triggering factor was the environmental debate; this debate is not uniform in the countries, but country-specific. It also differs among the sectors: especially car, chemical and packaging sectors are confronted with environmental debates on their products and the application of LCA is one appropriate way to respond to these claims.

According to our case-studies, policy as a driver for LCA is not important in Sweden and Italy, but possesses some influence in Germany and Switzerland. This is especially valid for the German waste legislation (packaging decree and Waste Management and Product Recycling Act [KrW-/AbfG]).

We recognised that clear bottom-up or top-down patterns do not exist: the Swedish and most German case-study-companies are examples of a bottom-up approach of LCA, whereas some Swiss and all Italian[5] companies introduced LCA top-down. In any case, the type of approach in the habitualisation stage is **not** a determining factor for success or failure within the whole institutionalisation process of LCA.

On the contrary, looking at these different factors, the importance of actors has to be highlighted[6]. The initiators are the people who push the whole process and this push is absolutely essential. They need not to be the entrepreneurs, but they might be. They are the initial driving force.

The plurality of drivers for starting LCA-activities was also checked within the survey. Important drivers for LCAs in all countries are cost savings[7]. Other impor-

---

[5] The influence of the parent company has especially to be remembered for the Italian case-studies.

[6] We identified three crucial actors within the institutionalisation process, namely the initiator, the promoter (or entrepreneur) and the set of champions. Their relevance changes during the process phases. The initiator is the central driver in the first phase. The importance of the entrepreneur (who might be, but is not necessarily, the same person as the initiator) is central in the objectification phase. His presence and influence are crucial factors for the overall success of the whole process. Finally, in the sedimentation phase, when consensus is needed, the set of champions involved in the diffusion of the tool has to expand within different sectors of the company. The set of champions generally includes the entrepreneur, however this is not strictly necessary, since in the full sedimentation phase the tool "survives" even if the people change.

Other actors involved in the LCA activities are the commissioner (the one who really gives the money) and the LCA analyst(s), who physically carry out the study. However, their role is less important, as long as they do not coincide with the initiator and/or entrepreneur.

[7] However, it is interesting to notice that the role of cost-savings as a driver is perceived differently in the four countries: They are mentioned explicitly as a driver in Germany, Italy and Switzerland, but in Sweden cost-savings are indirectly perceived via the future

tant drivers are product specific environmental discussions and problems[8]. Another remarkable finding is the importance of R&D in Sweden which hints at the proactive orientation of Swedish companies. In all four countries, the direct influence of the application of LCA by competing companies is not perceived as a driver (although the monitoring of competitors becomes important in the next stage of objectification). Environmental legislation, i.e. political or legal pressure, is not important, especially in Sweden and Switzerland; however, in Germany environmental legislation is ranked close to the most important drivers. One might also conclude that a long-term and proactive orientation of companies supports the start of LCA because LCA is able to analyse and describe future problems and risks of products.

These results confirm in general Smith et al. (1998) who identified market and regulation as main factors determining the adoption of LCA. However, in contrast to Smith's report, we looked at the implementation patterns from a dynamic point of view and recognised different institutionalisation stages.

The semi-institutional development: the crucial stage of objectification

After the initial and successful adoption of LCA in the habitualisation stage, it is necessary to develop a consensus regarding the use of LCA. The development of such a consensus concerning the use of LCA is probably the most crucial stage before LCA is applied routinely, and is the main characteristic of the objectification stage. The institutionalisation theory introduces two different mechanisms which contribute to a consensus during this stage:

- Monitoring and information gathering and
- champions who promote LCA.

The case-studies showed that a monitoring of the competitors took place in all companies which still use LCA, i.e. in all successful cases. The ways of doing this might be different, as our case-studies show: Henkel for example started its LCA because Procter & Gamble as competitor carried out an LCA on the same subject. Fiat Auto was involved in the EUCAR-project and had therefore the chance to co-operate and monitor other automobile producers. The Italian company ABB-Sace, which is owned by Swedish ABB, was confronted with activities in Sweden and stimulated to carry out LCA in its company.

In all the successful case-studies, an entrepreneur with a clear and reasonable strategy exists. The entrepreneur should have a "mission", that means a high personal commitment to push forward the LCA-activities. But the existence of an entrepreneur is not sufficient. Even more important is his/her influential position within the company. Examples of successful companies are Fiat Auto where the

due to liabilities. This is a modest hint at Swedish companies being more proactively oriented than companies in the other countries.

[8] A specific Italian phenomenon is the huge influence of the international management of mother companies on the use of LCA.

entrepreneur is the Head of a Department appointed at the very same time for both the Environmental and Industrial Policies of the company and Henkel where the entrepreneur is the head of the Quality & Environment Department of the Washing and Cleaning Agents (W&CA) Division.

The existence and importance of an entrepreneur is a necessary, but not sufficient condition for the successful promotion of LCA. Also a good strategy with some appropriate objectives is necessary. The objectives at the objectification stages are

- to get support within the company,
- to create links within the company and
- to enlarge the set of champions involved.

The entrepreneur needs to be persuasive, he/she needs to provide good justification for introducing LCA within the company. This justification can differ, but missing or insufficient justification will result in at least an uncertain future for LCA. In all our case-studies, LCA failed if the justification was not sufficient.

An appropriate strategy, communication and justification might make the creation of a large set of champions easier. All these elements might contribute to get a mandate from top-management to go further and implement LCA within the organisation.

Looking on the personal component of actors, it becomes clear that the entrepreneur is the central figure. He/she has to push the LCA activities and to involve representatives from different departments. Communication and information have to be carried out very actively to organise working teams and to exchange experiences. Top-management should also be involved and become sensitive to the organisational problem and the new tool LCA. The creation of champions supplements the diffusion of LCA within the company. During this stage, the influence of external advisers and consultants decreases.

The full-institutional development: sedimentation

After a successful objectification, during the sedimentation stage, LCA becomes implemented. At the end of this stage, a formal structure for working with LCA exists. This means that the use of LCA has now become taken for granted, is routine, institutionalised, and survives even if the people in the company change.

The success of the implementation depends on the strategy of the entrepreneur and the company-internal-relationships. During a successful objectification, the entrepreneur should have gained the support of key persons and groups. The importance of opponents should be held in check. To secure this political situation, the entrepreneur and the champion have to demonstrate positive experiences with LCA and a reasonable cost-benefit-relationship using LCA.

Once again, different actors play the decisive role for success.

The other crucial factor for full institutionalisation of LCA are company-internal co-operation and communication channels. First of all, they are necessary to make LCA known within the company in order to obtain support on the path towards a consensus and spread out the tool in other departments. In all our successful case-studies, a good organisation for the communication of the LCA results could be observed. At Henkel, this organisation was formally established from the very beginning, as a cross-product division working group of 10-12 people, including marketing, R&D representatives, ecologists and biologists. This immediately guaranteed the communication between different departments and divisions. At Fiat, LCAs are co-ordinated by the Central Laboratories, which already provide other technical departments with the data on materials.

The role of LCA changes

The role of LCA changes over time along the institutionalisational stages. In many cases and especially in Germany, at the very beginning, LCA was expected to be used in marketing. However, state of the art LCA methodology (especially in the area of grouping and weighting) is not elaborated and codified enough to apply LCA for marketing. In all our case-studies, marketing expectations could not be satisfied. Either LCA then failed generally or the adoption became mainly used for company-internal purposes. In general, companies realised that results are too complex and sometimes disputable to be presented to the market. This is due to methodological problems with respect to impact assessment and to the issue of availability and quality of data. Nevertheless these do not exclude the possibility of LCA being used in the long-term as a marketing tool if (and only if) some codified procedures according inventory assumptions, data quality and impact assessment procedures are agreed to by then[9].

LCA is first applied for learning purposes; this is an important characteristic of the habitualisation stage. We have recognised that also producers of complex products (e.g. automobiles, electric or electronic products) start with the examinations of simple parts of these products; e.g. the selection of weights for washing machines, of pipes of vacuum cleaners etc. The view is on existing products, i.e. a retrospective environmental check.

During the following institutionalisation stages, the role of LCA changes. During the crucial objectification stage internal purposes, both retrospective (especially bottleneck identification) and prospective (R, D & D) orientations, are important. LCA still maintains a high learning value.

If the first habitualisation stage has been passed successfully, and if the LCA was motivated by external market drivers, we observed that the expectations of the marketing department were not satisfied, but the internal application became more and more important. However, this depends on two crucial factors: the openness of people to unexpected results and a good justification for LCA.

[9] However, LCA can be, and is actually used, to provide information to suppliers, customers and stakeholders. In fact, the survey showed that this is already one main application in Germany, Sweden and Switzerland and that the use of LCA for this purpose is expected to increase in the future in Italy as well.

Indeed, the shift from external expectations to an internal application of LCA only happens if the people within the company go through a learning process, which allows them to tackle the dissatisfaction of unsatisfied expectations (marketing) and appreciate highly enough the added value of LCA for other purposes (R&D, product development, learning, spin-offs, etc.)

Another crucial factor is a good justification for LCA. Whatever the application of LCA, if there is no good justification, the institutionalisation process will fail. The clearest example of is occurs when companies *perceive* that they have only limited influence on the rest of the chain. In this case, even if they feel they have learned a lot about their products, they will nevertheless feel that LCA is ineffective (and unnecessarily costly) in diminishing environmental impacts of their products. In all cases of this type we observed companies either developing alternative environmental assessment methodologies or fading out their use of LCA.

If all the hurdles of this stage are overcome, then a clear prospective and company-internal application is preferred. The LCA results might still be used for external information but not for direct marketing activities. The use of LCA can be diverse depending on circumstances within a company, but the applications have to follow a clear and precise target-orientation. The use is a more prospective one looking towards the future. Examples within our case-study companies are the applications at Fiat and Volvo to assess new materials and components for R, D & D of new automobile development. Also the German company Henkel uses LCA for checking new compounds which might be used in the future.

In addition to this, there is a clear trend among producers of complex products: they started their LCA-activities with simple products or compounds which were (relatively) easy to analyse; if the learning of the tool LCA was successful, they then considered more complex products and/or fitted "simple" LCA-studies together (several compounds for a final complex product), see also Figure 7.1 for the description of this process.

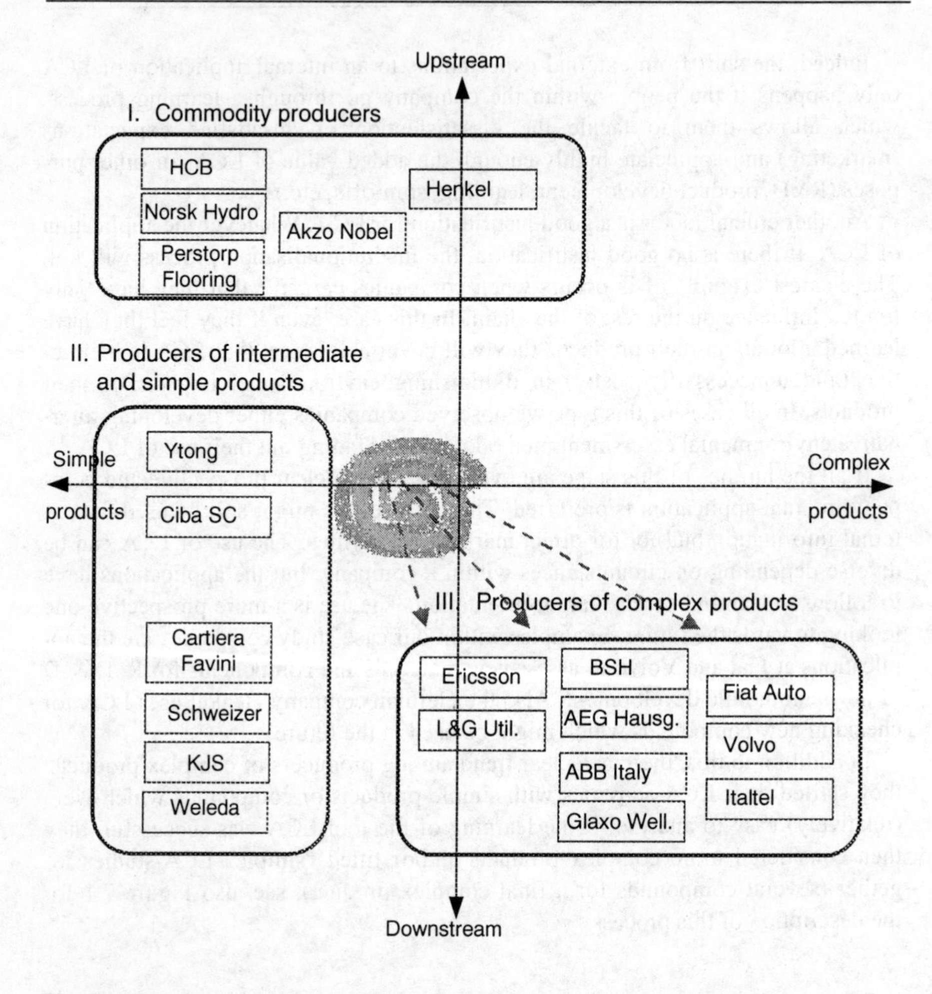

Fig. 7.1. Types of products and life-cycle position of case-study-companies

Altogether, the role of LCA changes from learning at the very beginning to doing after successful institutionalisation (see Table 7.1).

Table 7.1. The changing role of LCA along the institutionalisation stages

Pre-institutionalisation stage: habitualisation	Semi-institutionalisation stage: objectification	Full-institutionalisation stage: sedimentation
• marketing purposes in Germany and Cartiera Favini • mostly for learning • mostly for confirmation of status quo (retrospective) • simple products, compounds	• no marketing, but information to clients and stakeholders) • still high learning • mostly for internal purposes (R, D & D, bottleneck identification) • slow shift from retrospective to prospective use	• no marketing, but information to clients and stakeholders) • use is different, but precise • mostly for internal purposes (R, D & D, bottleneck identification) • prospective use • transformation of LCA results into action • complex products

Actions taken on base of LCA[10]

LCA is both intended to contribute to a company's knowledge and to reduce the environmental impact of its products. As pointed out above, the role of LCA changes during a successful implementation: there is a shift from learning to doing. Most of the companies examined in our case-studies are still in the learning process. Nevertheless, LCA supports some decision-making processes by providing additional information. It is clear that (today) LCA complements the "original" set of information while not automatically leading to direct consequences on the product.

Some of our case-study companies used LCA-results to affirm a planned decision. For example, BSH applied this instrument in the decision-process regarding which materials should be used for soap containers; the intended preferred material was confirmed as being environmentally advantageous. As already mentioned above, we conclude that during the habitualisation stage LCA is used especially for learning. Concrete decisions are more retrospectively oriented, namely to confirm existing considerations and/or intuitive feelings.

[10] See for the influence of economic aspects subsection 7.1.4.

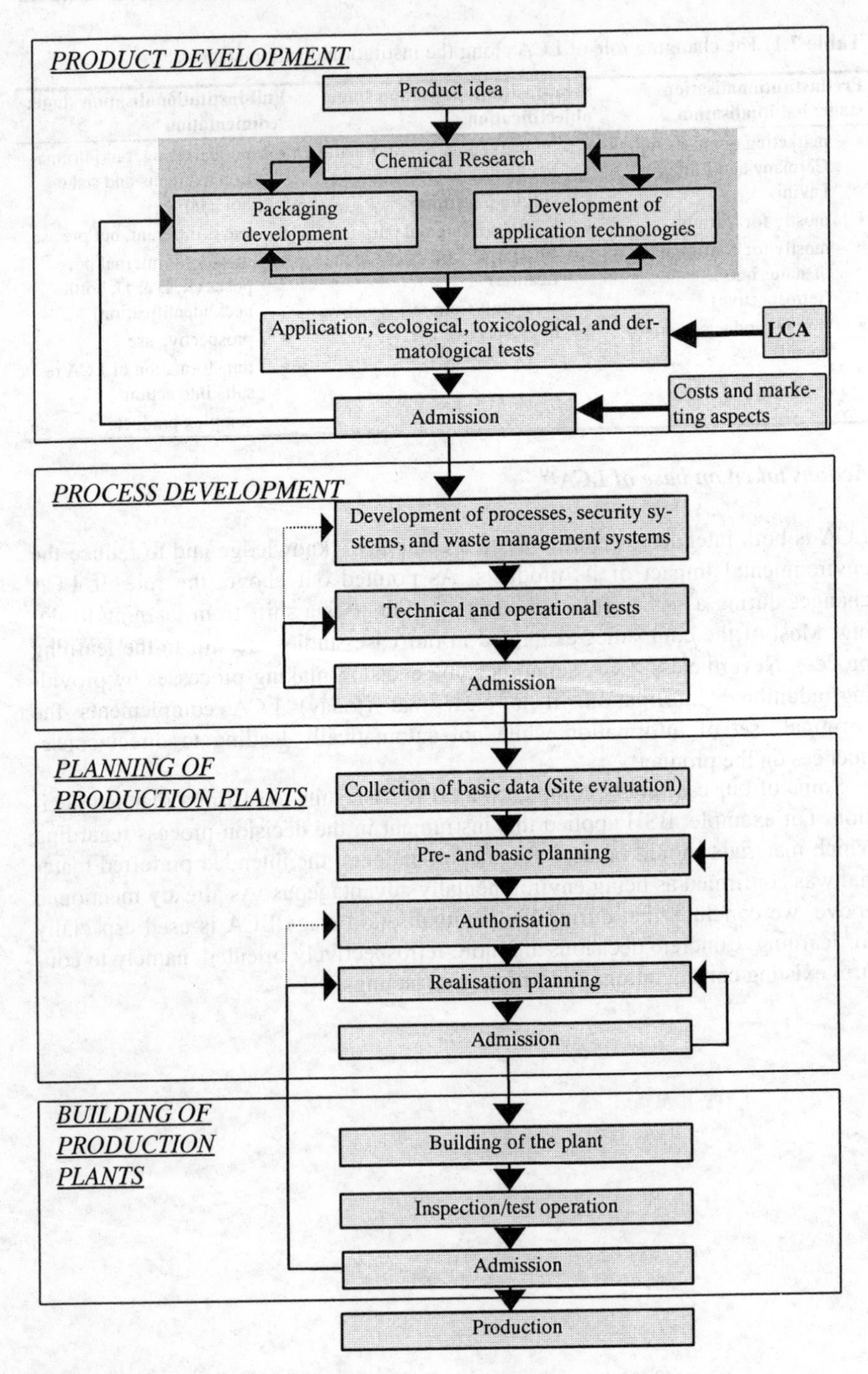

Fig. 7.2. From product development to production (see Henkel (n.y.), p. 18)

The case of Henkel demonstrates the changing role of LCA during the implementation stage. Whereas the first LCA was used for confirmation, LCAs are nowadays used prospectively for analysing substances and single products or to compare different product alternatives. LCAs supplement other instruments and do not substitute them. As a relatively complex instrument, LCAs are used only after other, simpler, screening tools have been applied. In the division of washing and cleansing agents the division management has given instructions to systematically carry out LCAs for every "important" product innovation. The complex LCA-results are summarised and made available to decision-makers in management and product development. They are inter alia presented in the form of so-called "collection of substance data". This data sheet encompasses the most important substance related information in a very condensed form. Selected life-cycle data (cradle to gate-LCA), basic data of the ecological and toxicological analysis and other relevant data are depicted. Henkel's entire process from product development to production is described in Figure 7.2. This figure shows that LCAs are used as an additional tool during the decision-making process. Each concrete decision is dependent on several time-place-related factors[11].

However, besides the direct link between LCA and a specific situation, LCA contributes to indirect actions, e.g. future R&D and product developments. In our case-studies, we found many hints that LCA stimulates R&D. For example Weleda, a company at the very beginning of LCA-activities, decided to consider the consequences of new more environmentally benign packaging within R&D[12]. Another example of future product development is Fiat Auto: As have many other car producers, FIAT has embraced the strategy of reducing the weight of its cars as much as possible with the goal of reducing fuel consumption. A possible way of doing this is by substituting conventional cast iron engine blocks with lighter ones made entirely from aluminium. However, the study demonstrated that the use of aluminium is environmentally positive only under certain conditions. The study concluded that, under the current state of the Italian recycling system and of the Italian market of secondary aluminium alloys, the use of aluminium on a large-scale would have to be taken into account with great care. This had significant implications on the strategy of the company. In fact, after the study, the decision was taken to slow down the substitution of cast iron engine with aluminium blocks and to apply it only to new models. In Cartiera Favini, the LCA encouraged new R&D activities in the alternative paper's raw materials supply.

The consideration of LCA within decision-making depends on the institutionalisation stage, good justification and, especially important of course, economic

[11] Henkel for example decided to put the result of the first LCA into action with no additional economic burdens; later when market prices changed, the decided action became too expensive and was therefore partly stopped.

[12] The LCA showed that any switch in packaging (plastics instead of aluminium) would need a different composition of the products. The reason for this, according to Weleda, is that currently no synthetic alternative is available with properties equal to those of the aluminium tube.

aspects. The economic aspects consist of different elements, for example the economic situation of the company, net benefits/burdens due to LCA-results.

Our case-studies showed that on the short term economic aspects always prevail. However, companies which use long-term planning, and where this also includes a long-term environmental orientation show a high acceptance for LCA-results. One example is again Fiat Auto; the environmental policy at Fiat Auto is defined in a long-term strategic framework program called "Project 21"[13]. This project was developed at the "Direzione Ambiente e Politiche Industriali" [DAPI] ("Department of Environment and Industrial Policies"), established in 1991. As a core tool for the wider concept of Life Cycle Management, LCA is explicitly included in "Project 21".

All told, we think that LCAs contribute to the systematic consideration of the environmental impact of products where routine or strategic decisions have to be made.

The impact on a company's culture

LCA is an innovative method:

- It pushes interested persons and actors towards adopting a new way of looking at business activities. The consumption of inputs like resources and the emissions into air, soil and water along the life-cycle confront a company with up- and downstream requirements. The former consideration of the environment from a company's gate-to-gate view is now supplemented by a cradle-to-grave view.
- New and innovative insights have been made in most of the case-studies. Most analysed companies declared that there were elements of surprise in the LCA-studies and that these were met with a high degree of satisfaction within R&D and technical departments.
- The institutionalisation of LCA, however, is only successful if a company's structure is able to adapt to the new demands. This is especially true for companies with a full institutionalisation of LCA. As presented above, Henkel integrated LCA in its product development process and Fiat Auto is going to systematically apply LCA in the integrated component development plan. Other examples exist also in companies which have still not reached a full institutionalisation. E.g. BSH has used LCA for its "Product-Environment-Examination", and ABB-Sace has appointed a new function for LCA.
- The implementation of the phenomenon LCA needs actors and especially an entrepreneur. He/she has to organise the process of institutionalisation of LCA. His/her role is crucial particularly in the Objectification stage. We hinted already that the support of advocacy groups is necessary to reacha full sedimentation.

---

[13] Project 21 takes its name from the Agenda 21 defined at the Rio Conference.

- LCA contributes to internal trainings: Ericsson, for example, uses a simplified version of LCA in internal staff trainings.
- LCA can stimulate co-operation and communication within and beyond the whole company, e.g. in Weleda (encouragement of LCA-application in Weleda's French and Swiss companies) and AEG Hausgeräte (co-operation with Swedish "mother" Elektrolux). At FIAT, LCA activities are co-ordinated by the Department of Environmental and Industrial Policies, but they also include other companies of the whole group.

LCA as a tool for talks, decisions and actions?

As mentioned above, the role of LCA changes. During habitualisation, LCA is a tool which is applied primarily for learning and secondarily interest for confirmation. In section 2.1, we introduced a distinction between three arenas:

- Talk,
- decision and
- action.

These three different alternatives offer action possibilities possessed by each social system. Taking into account our case-studies, we made the following observations: In the habitualisation stage LCA is especially used for talks. The outcomes are more of a secondary interest and not necessarily to be used for decisions. The really important decision is an organisational one, namely to proceed and continue LCA-application within the company.

Along the path of a successful implementation of LCA, the role of LCA also changes: from learning to doing. This means that LCA is still used for talks, but its importance for decisions and actions increases. At the end of a successful institutionalisation process, both adequate decisions and ongoing actions are an immanent consequence of the application of this tool.

Table 7.2. Importance of arenas along the institutionalisation stages of LCA

Arena **Institutionalisation stage**	**Talk**	**Decision**	**Action**
Habitualisation stage	A	B	C
Objectification stage	A	B	B
Sedimentation stage	A	A	A

Explanation: A = high importance
B = medium importance
C = low importance

Table 7.2 summarises our ranking of the application of LCA within the three arenas.

7.1.3 Actors, tools and methods

Actors, champions and practitioners

We have already highlighted the relevance of the main actors for the institutionalisation process, that is the initiator, the entrepreneur and the set of champions. Beyond them, there is another class of actors which is obviously important, they are the LCA analysts and practitioners. Their presence and role are significant with respect to how LCA is actually carried out and to which degree it is actually integrated within the company.

The institutionalisation of LCA changes the relationships among different actors. In our case-studies we found that the crucial factor for a full integration of LCA in company business is to which extent (if ever) the LCA know-how is actually "internalised" within the company. LCA-beginners often call in external LCA-experts and consultants. The ways of doing this are different, but the principle is the same: to use external support for learning. This can be a positive triggering factor for starting LCA activities. However, afterwards, this is not a guarantee for a successful institutionalisation process. On the contrary, our case-studies show that some of the failures of LCA can be traced back to the dominance of external consultants: they carried out complete LCA-studies and the involvement and training of company-internal staff was insufficient. An example is Cartiera Favini: Here the LCA was proposed and carried out by external consultants. The process did not involve internal resources in learning and, as a result, the company remained unable to do a new LCA without external support. This is one of the reasons why Cartiera Favini has currently no plans to proceed with an LCA strategy. This situation also holds in the German company Ytong and the Swiss firm Ernst Schweizer.

This means that during the implementation of LCA, LCA know-how has to be internalised within the company. LCA-capacities are necessary to carry out internal LCA-studies. These capacities have to encompass:

- personal resources, especially LCA-practitioners and
- technical resources, especially a data-base and software[14].

The increase of internal resources and the decrease of external consultancy is a clear characteristic of a full institutionalisation. Once more, the importance of actors has to be emphasised: initiators and entrepreneurs have to look to spreading out the approach and to involving other departments and people in using LCA (practitioners).

---

[14] The crucial point is the availability of inventory data. This is usually achieved through the development of an internal data-bank. However, in some cases it is also achieved by shared data-banks (either through sectorial studies or through the Corporate group). As far as software is concerned, we observed that most of the companies either developed an internal software from the beginning or they used a commercial software for learning and then adapted it according to their needs.

Our survey showed a clear dominance of internal teams in all four countries: In Sweden, the percentage of LCA studies carried out internally amounts to 77% of LCA-companies. But in the other countries, we observed that between 40% and 50% of the companies involve external consultants. This suggests that the "internalisation" of LCA competencies within the firm increases with the wider use of LCA.

LCA and other environmental assessment tools

The development of the tool LCA is an answer to the challenge of environmental pollution and deterioration. This challenge is nowadays a task for the whole society and all actors have to look to find their contribution to pathways to sustainable development. As a consequence, a lot of companies have acted and now consider environmental profiles of their products. Obviously, different tools support this consideration.

The empirical knowledge of the application[15] of these tools is very restricted[16]. Our survey (see subsection 4.2.5.4) showed that different instruments are used in the context of environmental improvements of products, e.g. LCA, checklists, material and energy balances. The most frequently used instrument seems to be compliance/gap analysis with legislation.

These different tools which can be applied in the context of the environmental assessment of products have specific characteristics[17], opportunities and restrictions. Nevertheless, perhaps the most complex and sophisticated instrument is LCA. We recognised that several ways of adopting the LCA-method for company-internal purposes exist (see also below). One specific way taken by several companies was to develop assessment tools based on an LCA, but using a simplification of this method. Examples are Landis & Gyr (development of a tool called "Product-Eco-Performance"[18]) and Italtel (development of a methodology called "Eco Compatibility Analysis").

---

[15] This statement is also valid for the state of institutionalisation of these different tools in companies (see also section 7.3 "Recommendations").

[16] Since 1993, the university of Hannover carries out panel surveys. In 1995 and 1997, several questions examined the state of environmental business management. One question asked for the application of different environmental instruments; as a result it became clear that the instrument "quality system" dominates. Checklists are used by 1/3 of the answering companies. The application of eco-balancing was also asked for; however, the term did not clarify whether LCA or company environmental reporting was meant (cp. Steinle et al. 1998).

Another interesting report is the Environmental Business Barometer (Belz/ Strannegárd 1997).

[17] See Rubik/Teichert (1997), Wrisberg/Gameson (1998) and de Smet et al. (1996) for a short description and characterisation of different tools.

[18] The tool "Product-Eco-Performance" is as sort of checklist with environmental criteria such as: efficiency of material, hazardous substances, packaging, energy use, ability of

Altogether, environmental tools and especially LCAs support the decision-making processes by delivering environmentally-related information; this information supplements already existing information and enlarges (and improves) the basis for decision-making.

LCA-methods

The development of the LCA-method is characterised by considerable progress on the level of standardisation. Two international standards have been agreed upon, others are under preparation (see Box C).

Box C: Status of ISO-standards on LCA[19]

- ISO 14040 - "LCA - general principles and procedures": Agreed and published as international standard;
- ISO 14041 - "LCA - Goal and scope definition and inventory analysis": Agreed and published as international standard;
- ISO 14042 - "LCA - Life cycle impact assessment": Draft international standard [DIS], might pass during 2000 as international standard;
- ISO 14043 - "LCA - Life cycle interpretation": Draft international standard [DIS], might pass during 2000 as international standard;
- ISO 14047 - "LCA - Examples for the application of ISO 14042": Approved working item, might pass during 2001 as technical report (type 3);
- ISO 14048 - "LCA - Life cycle data documentation format": Working draft, might pass during 2001 as international standard;
- ISO 14049 - "LCA - Examples for the application of ISO 14041": Committee draft [CD], might pass during 2004 as technical report (type 3).

The ISO standards 14040 and 14041 contain a lot of prescriptions about what should be done in which way. In particular, there are strong requirements if the results of LCA are intended to be communicated to any third party. However, this ISO-definition encompasses a part of different Life Cycle Approaches only. As mentioned in section 2.2, in reality companies are applying LCA with a variety of different approaches (i.e. by carrying out simplified and/or incomplete LCAs, or developing alternative assessment methodologies within a Life Cycle Thinking

recycling and management of life cycle. These criteria are rated and added to an overall mark for each product, which describes the environmental performance of the product. The product designer is obliged to control his design process with this tool. He has to reach with his product a minimum of points and therefore has a clearly described goal for ecological product design. The fact that several criteria in the PEP are based on information of LCAs carried out in Electrowatt or elsewhere show the importance of this tool, even if it is not used continuously.

[19] Information based on Annual Report for 1998 of ISO/TC 207 (ISO/TC 207 N 302).

approach, etc.). Figure 7.3 describes these approaches as a continuum between two poles: from Life Cycle Thinking (LCT) to full LCA.

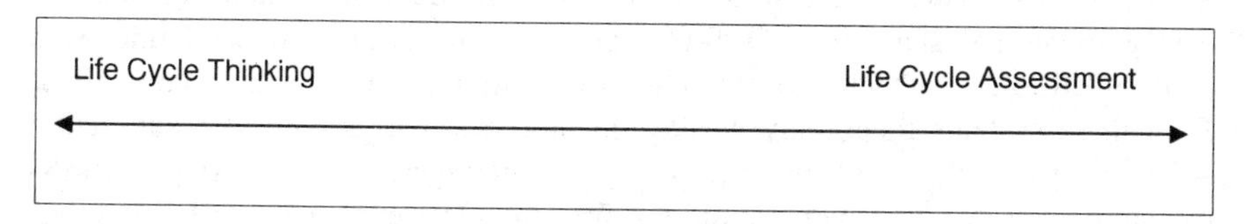

Fig. 7.3. Life Cycle Approaches (Source: Allen 1997, p. 13).

Our twenty case-study companies intended to apply LCA for company-internal purposes. As mentioned above, all companies which started LCA-activities with regard to marketing claims did not continue this pursuit. In addition to this, companies knew the sophisticated ISO-standards and used them as an orientation, but their way of carrying out LCA was more streamlined and is adapted to business-internal purposes. There are several ways of doing this:

- restriction to an inventory without any further impact assessment or interpretation;
- restriction to some key indicators (example: consumption of energy, emissions contributing to global warming);
- application of "simple" weighting methods like net value method (example: BSH) or EPS (examples: Volvo, Ericsson, Perstorp);
- developing alternative methods based in some way on LCA information (examples: Landis & Gyr, Italtel).

Altogether, we identified a clear trend and also a need for simplified LCA-techniques. During the implementation process, the type and character of LCAs change; reaching a higher institutionalisation stage (and using LCA for internal purposes), the method was adapted to the application areas; it became simpler, leaner and more streamlined.

Nevertheless, it has to be mentioned that companies do not want to stop or ignore the ISO-activities. The main reason is that they consider these activities to be an excellent tool for communication among and within companies and also provide a good reference point in dialogue with stakeholders. But in addition to this a general demand for streamlined LCA-techniques exists[20].

[20] It is important to notice, that ISO is seriously addressing the question of simplification rules. In particular, recognising that the issues of simplification might be sector-specific, in a recent workshop of CEN it was proposed to establish working groups on LCA focusing on particular industry sectors. Part of their work should be also to establish simplification rules.

The importance of LCAs in environmental management systems (EMS)

Environmental management systems have been introduced in many companies. The international standard ISO 14001 and the European EMAS-scheme have contributed to a systematic consideration of environmental aspects in companies. Both these systems focus only weakly on products. Our survey (see subsection 4.2.1.2) showed that, with the exception of Switzerland, already existing EMSs can support LCAs. However, we did not find any clear evidence that LCA activities support the implementation of EMSs.

The Federal Environmental Agency (UBA 1998) carried out an investigation of German companies registered according to EMAS. They concluded that only a small minority (17%) of the answering companies indicated that they would use EMAS because they want to discover product or process innovations. This was also confirmed by an overview prepared by Elsener et al. (1998). However, FEU (1998, p. 198ff.) evaluated the EMAS scheme by interviewing selected German companies; they found that 70% of the interviewed companies describe their products in the environmental statement and that 2/3 indicated using environmental criteria in product development and life-cycle consideration[21]. The planned revision of the European EMAS-scheme might strengthen the focus on products. Several proposals exist (cp. Schmidt 1998, Rubik 1998b), however, the decisions are still under discussion and will not be passed before 1999.

Nevertheless, Jensen et al. (1998) report on a survey which was carried out by SustainAbility late in 1996[22]. The respondents expected as a likely trend that "LCA will be seen as an integral part of the environmental management tool-kit, but will also find new applications in areas such as corporate strategy" (p. 28).

7.1.4 The potentials of LCA – some outcomes

LCA offers several potential benefits. Often it is argued that a two-fold dividend, from both economic and ecological revenues, could be realised by using environmental management tools (cp. for example Gege 1997). However, clear proof of this hypothesis is complicated. Some studies have assessed the potential benefits of LCA:

---

[21] This information is based on statements which could not be verified; therefore the indicated numbers should be interpreted with caution.

[22] The survey was carried out in summer 1996 by SustainAbility on behalf of the European Environmental Agency. SustainAbility interviewed 43 LCA "opinion leaders" from different stakeholder groups all over Europe: companies. research institutes, eco-labelling boards, NGOs, practitioners, financial institutions. The number of companies interviewed was 13: they were selected from among the most active in the LCA field (e.g. Dow, Fiat, Procter & Gamble, Body Shop). More information on the survey is available on the EEA web site.

- Rubik/Grotz/Scholl (1996) conducted a study aimed at assessing the environmental benefits of an LCA application at company level. Their research focused on the description and evaluation of eight case studies from different industrial branches which analysed whether LCA-studies contributed to reductions in environmental burdens.
- In a study on behalf of DG III from the European Commission, Smith et al. (1998) interviewed about 70 companies from six industrial sectors (aluminium, chemicals, building materials, personal products, electronic goods and automobiles) about their adoption of Life Cycle Approaches. They examined the impact of LCA on innovation, competition and trade; according to their research, the impact is determined by the technological and market environment.

Our research revealed some hints of LCA potentials:

The Impact

We got some material on the impact of LCAs:

- Results of LCA and economic consequences: The Henkel case-study showed that of two resources assessed by the LCA-study, one was shown to be environmentally advantageous; however, the decision to market this product with the preferred resources was not completely realised because the advantageous resource became, due to a massive cut in prices, too expensive.

  At FIAT the manufacturing extra-cost for the environmental improvement of the car and its components is in principle allowed. However, in messy situations, economic arguments always prevail.

  In contrast to the hints from the case-studies, our survey found out that the implementation of measures proposed as a result of LCAs are not an important obstacle to a wider use of LCA (see subsection 4.2.4.1): only 10-18% of the companies in the four countries indicated this as a main obstacle.
- Impact on investment: Our survey showed that most companies believe that the benefits of LCA are long-term (see subsection 4.2.4.2). This means that any investment in LCA-techniques does not bring an immediate return-on-investment. On the other hand, LCA-results might influence the investments in new technologies of a company. An example is BSH: So far the soap container in BSH washing machines has been made from steel. In 1994, in the course of the development of a new top-loader washing machine model, the product development department of the laundry product area thought of replacing the steel container with a polypropylene [PP] variant. The switch from steel to PP was beneficial for economic and technical reasons, but required completely different production processes and thus meant an important and hardly reversible change in the basic technology. Therefore, it was decided to carry out an LCA on these two different materials used for soap containers before any change of the technology took place. At the end of the LCA-study it was decided to apply the environmentally more benign material, and the investment policy was correspondingly oriented.

- Impact on environmental issues: We collected a lot of material from the case-studies showing that reductions of environmental burdens have been stimulated by LCA. They encompass the improvement of environmental bottlenecks (gradual improvements) to system changes (see example of BSH on PP as material used for soap containers).

These findings are mostly anecdotal ones and should not be generalised.

Innovation[23]

LCAs support innovation. We collected some material with regard to the innovation potential of LCA:

- *Organisational innovation:* This is the most important innovative impact of an LCA. According to our research, the implementation of LCA requires the adaptation of the structure of a company to the new challenges. New arrangements have to be agreed on. Companies in which LCAs have been institutionalised adapted their organisation by creating an appropriately modified structure. The first example was Henkel which integrated LCA in the formalised product development process (see Figure 7.2). Another example is FIAT, which (in 1998) was going to integrate LCA into the "Integrated Development Plan of car Components"; in this plan a new specification has been introduced for designers, that is the environmental impact of materials and of project alternatives. It is with this in mind that we predict that environmental matters and LCA results will become an everyday tool for designers and people involved in technical choices regarding materials and car components.
- *Product innovation:* Some, but not all case-study companies declared that LCA will also in the future be considered in R,D & D. LCA can stimulate product innovation, but the evidence in the case-studies is not so clear and depends especially on the institutionalisation stage. Broadly speaking, the higher the institutionalisation level, the closer the connection of LCA to product innovation.

  The survey (see subsection 4.2.2.3) showed that the general application of LCA for the assessment of all new products of a company is the exception; if at all, LCA is used for checking some new products. However, this can be explained by the fact that most companies are still in the habitualisation stage and only a few are in the sedimentation stage.
- *Process innovation:* Changes in products might imply changes in the technologies and processes which produce them. An example has been presented above:

[23] Berkhout (1996) and Smith et al. (1998) followed Dosi's (1988) characterisation of innovations and listed three main factors which generate innovation: technological opportunities, market demand and appropriability. Based on these factors, they analysed the influence of LCAs. From their interviews, they observe: "Few firms interviewed (...) could identify clear-cut examples of innovative or competitive impacts of the application of LCA" (Smith et. al. 1998, p. 43). The focus of this book is not oriented towards an analysis of competitive advantages due to the application of LCA. As a consequence, we are not able to check Smith's findings.

BSH's decision to substitute steel as a material for soap containers by PP. Another example is Perstorp which decided to introduce a new production method.

Integration of LCA in decision-making

The integration of LCA in business decision-making was also stressed by Smith et al. (1998, p. 48) a very small number of firms have actually integrated LCA-based-tools into their business decision-making. Many say that they are planning to do so without yet having a clear strategy for doing so". Our research shows that more than 50% of the case-study companies told us that they are already using or willing to use LCA as an additional support in decision-making procedures.

Realisation of environmental potentials of LCA

As mentioned above, Rubik/Grotz/Scholl (1996) recognised that the realisation of environmental benefits of LCA depends on several factors which influence decision-making in general[24].

This volume emphasises the importance of the organisational implementation of LCA. As mentioned above, a good justification of LCA is necessary for the successful institutionalisation of LCA. If this happens, then the contexts and circumstances of each decision influence the concrete decisions and herewith the realisation of potential environmental benefits. In other words: The more appropriate the justification of an LCA, the better the chances for a realisation of potential environmental gains.

Double dividend

The recent debate on eco-efficiency[25] and double dividend argues for both economic and ecological benefits following the application of environmental management tools. In some cases they might be easy to realise, in others less. In particular, eco-efficiency examples show that win-win situations exist when there is a shift from the production of goods to the sale of services. At the same time, a core aspect of eco-efficiency is a systemic (life-cycle) viewpoint with respect to both processes and products.

We think that this link exists potentially. However, we did not find any clear evidence that LCA supported eco-efficiency strategies. On the contrary, the application of LCA for a shift from products to services is the one ranked lowest in the survey. The use of LCAs for a radical change in a product life-cycle is also ranked low. Our interpretation of this is that it is simply too early. Such eco-efficiency

---

[24] See also Scholl (1999).

[25] See for example OECD (1998a) and De Simone/Popff (1998).

strategies require changes in consumption patterns and require time. However, in principle the link is possible and should be further investigated (see section 6.3).

What is valid is that LCA sometimes indicated cost-saving opportunities indirectly through spin-offs.

7.1.5 Are there any specific application patterns?

We discuss application patterns with regard to several criteria[26]:

- Countries,
- sectors and
- size.

Countries

This book (see chapter 3) that the level of LCA-activities is high in Sweden and Switzerland (criterion: GDP), very modest in Italy and well introduced in Germany. However, there seems to be an increasing trend in Italy and a decreasing one in Switzerland. This is confirmed by the 20 case-studies, which clearly showed that the degree of institutionalisation of LCA within the four countries differs.

According to the case-studies, especially Swedish companies are very busy with LCA and some of them have already reached the objectification stage (and Volvo the sedimentation stage). Several findings from the survey hint at a more proactive behaviour of Swedish companies and an openness to Life Cycle Approaches.

In contrast to this, the application of LCA in most Swiss case-study companies seems to be decreasing. We interpret this as a type of saturation in Switzerland.

This picture might depend on the accidental selection of companies within the sample of case-studies, but the results of the survey endorse this impression. As presented in subsection 4.2.4.4, Swiss companies are more sceptical about the future application of LCA within their companies than companies in the other countries.

If this description of the situation in the four countries is not misleading, then either the institutionalisation patterns differ or the national context is different.

Within our case-study research, we did not get a hint of there being any specifically national institutionalisation patterns. Or the other way round: implementation patterns do not differ between countries; the country is *not* a determining factor for the institutionalisation process.

On the other hand, we believe that the national contexts in the four countries do have to be distinguished. As described in sections 6.1 and 6.2, different national

[26] Other interesting criteria might be the position in the life cycle, the external environmental pressure and the nature of competition of the selected case-studies. However, it was not possible to consider these criteria due to budgetary restrictions.

LCA-cultures exist. An example of this is the different nature of relationship between business and policy: It is more co-operative and supportive in Sweden and in Switzerland, there is a modest awareness and limited support in Italy, and conflict and debated support characterise in Germany. However, this does not explain the differences between Sweden and Switzerland. We think that several factors contribute to the Swiss situation:

- During the last years, the attention of Swiss companies moved towards environmental management systems (30-40 sites are registered per year). As a consequence the attention on LCA decreased and became saturated.
- In contrast to Sweden, no large co-operative studies in the fields of eco-design or LCA-studies exist; the previous large and important LCA-studies are nowadays "state-of-the-art".
- Another reason might be that Swiss companies had great expectations about the application of LCA for marketing. These were disappointed. As a consequence, we suppose that Swiss companies tend to be left in a more waiting position.
- Another reason for the different situation in Sweden and in Switzerland has to be found in the larger use of LCA in the public sector in Switzerland which makes LCA more a tool for public planning and decision making and not only a company-management tool.

In the survey, most companies answered that they have a retrospective view on products and/or apply LCAs only for some new products. Especially Swedish companies seem to apply LCA more to new products than German and Swiss companies (see subsection 4.2.2.3). These answers might be interpreted as an indicator of a more learning and starting off attitude towards LCA in Germany and Switzerland. Otherwise a more arranging approach would have been taken.

Finally, the type of initial approach can be different from country to country (clear examples: top-down in Italy and bottom-up in Sweden, with Germany and Switzerland somewhere in between). This may well be connected with specific different local management cultures. However, we found that the type of approach in the habitualisation stage has no real influence on the success or failure of the whole LCA institutionalisation process.

Sectors and life-cycle position

Another important factor which might influence the application patterns of LCA is the sector and also the life-cycle position to which a company belongs. According to Smith et al. (1998), the position in the life-cycle is the *key* determinant of the approach taken to LCA adoption because the technological and market contexts differ along the life-cycle. They distinguish two typical patterns of LCA-approaches, namely:

- *Commodity producers:* Their approach "is externally-oriented, top-down, and is limited to the collaborative production of inventory data on a segment of product chain activities. This approach is typical in upstream, commodity-producing sectors (...)" (Smith et al. 1998, p. 41).

- *Complex final goods producers:* Their approach "is internally-oriented, bottom-up and includes studies across wider spans of the product chain which may include an evaluation component. This approach is typical for downstream assemblers of complex final products (...)" (Smith et al. 1998, p. 41).

Between the two basic types, a third type of producer of intermediate and simple products has been introduced[27]. The distinction into these basic types are simplifications, but they illustrate different patterns. Therefore we tried to test them in our report. We organised our case-study companies in the similar three groups marked in Figure 7.1.

An important distinction between Smith's and our analysis is the approach: Whereas Smith et al. provided a *static* picture of the adoption of LCA, we developed a *dynamic* picture. Because of these different research approaches, we can not verify their adoption patterns. Moreover we only performed 20 case-studies and most companies belonged to the group of downstream companies.

However, our feeling is that the adoption of LCA is much more dependent on other parameters than on the life-cycle position (and its characteristics as indicated by Smith). To be more precise: the business-internal adoption of LCA follows a specific logic of institutionalisation. Within this process, different characteristic elements influence the whole development. However, the type of Life Cycle Approach does *not* seem to be influenced by the life-cycle position of the company: e.g. Henkel (upstream), Volvo and Fiat (downstream and complex) are all in the sedimentation stage and use all LCA for rather internal applications.

In particular, we observed that no company is currently using LCA for marketing. Moreover, the introduction of LCA in companies producing complex

[27] Smith et al. (1998, p. 41) condensed the patterns in the following table:

	I. Commodity products	**II. Intermediate and simple products**	**III. Complex products**
Orientation of life cycle activities	External • Marketing • Policy Process	External and internal	Internal • Decision-support
Study structure	Collaborative	Collaborative and independent	Independent
Study practitioner	Third party and in-house studies	In-house and third party studies	In-house studies
Study type	Cradle-to-gate life cycle inventory (LCI) studies	Scoping total product cycle LCI studies	Total product cycle LCI/LCA studies
Data origin	Internally-generated	Internally- and externally-generated (upstream)	Externally-generated (upstream and downstream)
Evaluation	No evaluation	Limited evaluation	Some evaluation
Adoption process	Top-down	Variable	Bottom-up
Collaboration	On data	On data	On methodology

products can equally happen top-down or bottom-up (and this has no influence on the fulfilment of the institutionalisation process of LCA).

However, this does not exclude marketing people from being involved in the institutionalisation process of LCA. For example, at Ericsson, it was a product specialist with some environmental responsibilities who was very active and who initiated corporate-wide LCA activities, although he was positioned within one of the business areas.

This also does not rule out that LCA can be used for market communication with industrial clients (i.e. not advertising to end consumers). This is the case of most Swedish case-study companies, i.e. Perstorsp, Akzo Nobel, Norsk Hydro and Volvo. Moreover, the survey shows that information for customers and stakeholders is one main application of LCA in all countries.

Size

The size of a company seems to be an important factor for both starting an LCA and eventually fully integrating it in business activities. The inventory of LCA-studies showed that SMEs rarely carried out LCA-studies (see subsection 3.1.5). Looking at the case-studies it is interesting to notice that no SME company is in the objectification or sedimentation stage.

What might be the reasons? Beside the general barriers identified (see next section), some barriers are especially relevant for SMEs, namely:

- *Lack of know-how:* Often it is completely missing; other influences derive from a missing and/or "lost" entrepreneur. Relying completely on external consultants rarely helps companies go through a real learning process (which we observed to be very important for the full adoption of LCA).
- *Missing justification of LCA:* The possibilities of SMEs to influence up- and downstream activities are very restricted and therefore the chance to influence suppliers and clients is modest.
- *Costs dominate, benefits are modest:* Especially SMEs do not have the financial capacities to invest in the learning phase which is crucial for understanding and applying LCA. It has to be remembered that the benefits of LCAs are perceived as long-term ones.

We believe that the institutionalisation process of LCA is influenced by the size of a company. We think that SMEs have more problems organising the institutionalisation process in a fruitful way.

7.1.6 What are the barriers and opportunities?

We collected some information with regard to barriers and opportunities for a wider and more appropriate application of LCA.

Barriers for the future application of LCA

Our application of the institutionalisation theory revealed some crucial points which must be avoided if the implementation of LCA is to succeed:

- *Insufficient justification:* Any LCA-culture is embedded in a context. If LCA does not fulfil the expectations for a reasonable outcome (=justification) of the relevant actors, then it fails. Examples are Ciba Speciality Chemicals and KJS: The perceived influence of the company applying LCA-results was regarded as very restricted. As a consequence LCA-activities were regarded as not useful.
- *Lack of entrepreneur:* We have several times hinted at the importance of the entrepreneur. His/her existence is a must, otherwise the whole process of LCA does not advance. This might explain the insufficient implementation of LCA in the case of Ytong.
- *De-central position of entrepreneur:* Beside the existence of an entrepreneur, he/she should have some influence (not necessarily formal power). If this does not happen and/or does not hold with time, the institutionalisation process might fail (e.g. at Cartiera Favini).
- *Disappointment of marketing:* The application of LCA for marketing and business-external communication/information is unsuccessful at the moment. It is remarkable that within our case-study sample, often the marketing department joined (or initiated) the process of LCA. But their hopes and expectations were disappointed because the results of LCAs are not easy to use for marketing; only in some cases have LCA-results been used, but these were part of environmental reports or declarations of companies (example: BSH environmental report) and not as a marketing tool.
- *Missing openness to unexpected results:* If LCA did not fulfil the expectations of companies, the adoption of LCA was stopped (e.g. Ytong, Schweizer). In contrast, if unexpected results are accepted, a learning process is triggered and the institutionalisation process can continue (typical example is the shift from expected marketing purposes towards internal applications).
- *Lack of company-internal support and mandate:* The implementation of LCA is a time-consuming process which needs the clear support of advocacy groups, an involvement of management and for its full institutionalisation a mandate from top-management.
- *Influence of opponents:* Innovation causes new company-internal arrangements. Normally, the structure changes and not everybody is happy with this. If opposition towards the innovation grows, the implementation process of LCA will also be affected. Afterwards, it is a question of influence and power.

These above-mentioned barriers are more organisational ones. In addition to these, we identified in the survey some other barriers which are more methodological and/or technical ones (see subsections 4.2.3.3-4):

- *Disputes and controversies over results:* This point was regarded as the main obstacle for a wider application of LCA within all four countries which we examined. This is the main reason why LCA is not used for direct marketing.
- *Methodological problems, especially collection and quality of data, difficulties with assessment and interpretation and definition of system boundaries:* Except within Swiss companies, about 40% of the companies in the other three countries indicated that methodological difficulties are an obstacle.
- *Difficulties to communicate the results:* As elaborated above, the complexity of the results is perceived as a barrier to wider LCA use. This point was listed by 10-22% of the companies.

As an economic factor, the costs of carrying-out LCAs were also regarded as a barrier in the survey. Of course, this is particularly important for SMEs.

Altogether, these barriers confirm and enlarge on Smith's results[28].

Opportunities for the future application of LCA

Beside barriers, opportunities for a more intensive application of LCA exist. Within our case-studies we got some evidence that the following factors are opportunities:

- Environmental awareness and environmental management: These are necessary conditions for the willingness to consider gate-to-gate external impact.
- Realisation of double dividends / eco-efficiency potentials.
- Up- and/or downstream co-operation and networking.
- Participation in collaborative studies.
- Learning attitude of company and openness to unexpected results.

Of course, avoiding the barriers which have been listed above are opportunities for LCA.

These results confirm and enlarge on the results of barriers and opportunities which Rubik/Grotz/Scholl (1996, p. 104) identified and which are presented below in Table 7.3. The factors which have been verified by this book are printed in *italics* type.

[28] They identified costs, methodological indeterminacy, access to data, focus of studies and communication of results as the main barriers (Smith et al. 1998, p. 38ff.)

Table 7.3. Supporting and hindering factors for the application of LCA (Source: Rubik/Grotz/Scholl 1996, p. 104, own translation)[29]

Factor	Carrying out an LCA-study	Realisation of results of LCA-study
Costs and benefits	+ *Expected cost increase due to modified political framework* + success of market due to a positive image of the company + *demand of customers ('ecology pull', safe existence)* + availability of subsidies + *large costs of an LCA*	+ *cost-reduction due to saving of resources or recycling* + expected profits due to positioning in an ecological market niche + possibility of shifting costs onto the final customer + possibility of internal subsidies – insecure market success
Information/ communication	+ *environment-orientation as part of the philosophy of the company and open information policy* + *pressure of competition (publication of an LCA by a competitor)* – not enough internal and external co-operation and communication	+ support of the reliability of environmental advertising + communicative application of results (e.g. check lists/ eco-labels) + lose relationship to customers and suppliers – insufficient co-operation between practitioners and applicants of LCA – high complexity of results
Organisation	+ fixing the operational procedure (for example in own environmental department) + standard software system – missing banks with up- and down-stream information	+ participation of the employees in working groups – *insufficient integration of the concerned departments* – inadequate sales organisation
Technical and technological aspects	+ *low technical complexity of products* – *high technical complexity of products*	+ *capacity to realise suggestions* + realisation of the suggestion means expansion of technical know-how – existing production technologies
Other factors	+ low variety of products – *high complexity of method* – *no standardised weighting/valuation methods for LCA* – preparing LCA is too time consuming	+ LCA-based improvement proposals

Explanation: + = supporting factor
– = hindering factor

Future use of LCA

A clear prognosis on the future of LCA depends on the awareness of the changing role of LCA which implies a shift from learning to doing. This means that the implementation of LCA has to be regarded as an investment which encompasses personal and technical resources. If this investment which seems to be high at the be-

[29] The factors which have been verified by our study are printed in italics type.

ginning is structured and implemented in an appropriate way considering the factors accompanying the institutionalisation process, then LCA might be used as a routine tool. But of course, support is necessary; we present some proposals on these measures in section 7.3.

7.1.7 Expectations on LCA-applications and policy

The history and the practices of LCA do not restrict the application of LCA exclusively to business. The ISO-standard 14040 lists several application areas of LCA, namely:

- Product development and improvement,
- strategic planning,
- public policy making,
- marketing and
- other.

However, the standard itself is focused to the method of LCA; it does not consider the application fields of LCA. Current clients of LCA are especially business, but also governments and governmental agencies commission LCA-studies. The role of LCA for public policy making is not clear. There are considerably different views on it. An example is the German discussion on two packaging LCA-studies carried out recently on behalf of the Federal Environmental Agency [UBA] (see section 6.1). Beside the state, also different company-external stakeholders are involved in the development and application of LCA; they have hopes and fears about LCA.

All in all, the use of LCA within business is observed by stakeholders and environmental policy makers. Especially environmental policy has a specific view on business' application of LCA.

Why do policy makers act?

The attention on product-related environmental issues is a relative new one. The interest to assess the environmental impact of products and processes has been more intense in Europe than elsewhere, especially in northern countries. Policy makers started to refer to life-cycle concepts or cradle to grave approaches already at the beginning of 1980s. The activities of European regulators and regulatory agencies have also been a major factor influencing both the pace and the nature of LCA development in Europe.

The general principles of environmental policy in the four surveyed countries are going along with the model of sustainability. There is a clear shift from traditional command and control policies to market-based instruments and voluntary agreements in the form of co-operation between industry and public authorities. 'Polluter Pays Principle' (PPP) together with and 'Extended producer responsibility' are the main principles of the new environmental policies of the 1990s. However, the application of these common principles at the European level is taking different forms and patterns of implementation, according to the country considered. Northern countries have been keener on a more systematic and comprehensive application of these principles (including the 'precautionary principle'). Southern countries tended towards a soft application of the same (i.e. shared responsibility instead of extended European responsibility and soft environmental charging systems).

A more widespread application of the life-cycle concept in business may improve the public policy process by providing more comprehensive information to decision-makers. Because of these reasons, policy and other stakeholders are involved in Life Cycle Approaches.

Policy perspective: How should business apply LCA and how is it applied?

Policy has some expectations with regard to the application of LCA by business (see subsection 6.2.4). Does business fulfil them? Are there any gaps? In Table 7.4, we compare the expectations of policy with our findings and conclusions.

LCA is expected to contribute to the development of *eco-efficiency concepts* and their application. The relevance of LCA for these concepts depends on the role of LCA within a company and with its implementation. Companies which have just started LCA-activities use them for learning. And learning does not directly imply any relevance for eco-efficiency concepts. However, in the long run, companies which are applying LCA in a prospective sense support eco-efficiency concepts just by practising LCA.

Table 7.4. Expectations of policy on business' use of LCA and current practices

Expectation of policy towards business	Current practices of business
Supporting 'Eco-efficiency' concepts / application	• Learning from LCA: not important in the short term • Doing LCA: important in the long run
Tool for developing indicators for measuring performance	Not yet in existance
Potential application for strategic decisions	Depending on institutionalisation level
Identification of improvements potentials	Yes, in all countries
Raising environmental awareness	Yes, examples are LCT, use of LCA for learning (and sometimes for doing)
Identification of environmental priorities	Seldom for comparisons between different products of the same group; not for comparisons between different product groups
Supporting environmental objectives in companies	At the beginning

The development of *environmental performance* indicators started some years ago. Also ISO included this as an objective of standardisation[30]. According to the case-studies, the results of LCA-studies are still not used for the development of indicators useable for measuring environmental performance.

The identification and reduction of *environmental bottlenecks and potentials* is one important expectation which is clearly fulfilled. The survey shows that the application of LCA for this purpose dominates in all four countries (see subsection 4.2.2.1).

LCA is a tool which *sensitises* its applicants per definition for upstream and downstream environmental aspects. A high learning rate can be expected. This was verified by the survey; a lot of companies were confronted with new insights and surprises (see subsection 4.2.4.3).

The *identification of environmental priorities* is the task of each company. These might be fixed in an environmental report or declaration (cp. Fichter 1998). Environmental priorities in these reports, declarations etc. may refer to company external environmental discussions (e.g. to a national environmental plan, discussion with stakeholders). In our case-studies, we did get hints that environmental priorities need to be defined within each exemplary study. E.g. in this way BSH fixed and prioritised some weights for the regarded impact categories. We found no evidence of comparisons being made among different product groups.

The support given by fixing *environmental priorities* is, therefore, according to our findings something which occurs at the very beginning of an LCA.

Altogether, one gets the impression that policy still has an unclear picture and restricted knowledge of the business' use of LCA. Expectations and hopes of policy are huge, but these are not in the line with the response of business. Consequently, we think a mismatch between policy and business has already arisen or will arise.

Policy initiatives and activities

Governments and environmental agencies are active in the development of LCA (see subsection 6.2.5). There are several principles which refer to LCA in business: first the introduction of Life Cycle Thinking (LCT) in companies; second the help-for-self-help and good-example-strategy in order to achieve sustainable development in production and products; and finally, consumption. Within the spirit of sustainability all efforts in this field aim to close substance loops in the long run.

Whilst we identified many different measures, and actions and forms of support (see subsection 6.2.5), the measures which should be taken by policy to support business are still not well enough co-ordinated and structured in formal programmes. Nevertheless, there are a plethora of different measures and actions

[30] Within ISO-series 14030, the topic environmental performance indicators are going to be standardised.

which have been taken or considered by governments in the four countries. They have been involved either directly, in financing LCA to be used for their policy processes, or indirectly, in supporting the implementation of LCA in industry as an instrument for achieving sustainable development and in financing projects for methodological refinements. Worth mentioning is the absence of support for the use of LCA in SMEs. It seems that SMEs are looking increasingly for assistance in integrating LCA in their business management as an answer to both market and legislative pressures.

What does business expect from policy?

The expectations of business on policy measures in the arena of LCA are both unclear and vary between the countries. Companies are aware of the political framework and the interviewed case-study companies asked for integrated product policy actions to strengthen their environmental approaches. Some of these expectations are concrete and based on existing tools (support of official eco-labels, environmental orientation of public call for tenders). Some measures are more future oriented (e.g. the proposal by one company to establish a general product classification with classes determined by the environment-friendliness of products). One has to question whether these claims are in the "mainstream" of business' expectations. But they do represent some interests.

We found that different characteristic policy "styles" in the four countries we considered exist: the Swiss one (characterised by harmony and accepted support), the German one (characterised by conflict and debated support), the Italian one (characterised by modest awareness and very limited support)[31] and the Swedish one (characterised by harmony and accepted support).

Policy instruments and the view of business

In general, with the only exception of product and packaging take-back systems, the majority of companies tend to prefer voluntary instruments, namely eco-auditing and certification schemes. The survey showed that product standards and covenant/sector codes of practice are also very highly ranked in some countries.

Companies accept the concept of LCT within integrated product policy measures, but they strongly mistrust the use of LCA as a means of command and control. They think that this is inappropriate at the moment because of methodological issues (results are disputable). Moreover, they think that such a use may hinder rather than foster the internal innovative driving forces for starting LCA activities in companies. LCA would then be used "just as another regulatory requirement".

[31] However, some months ago, the situation changed when it was decided to found an Italian association for LCA by the end of 1998. Its members are researchers, employees of different Italian companies and representatives of the Ministry of the Environment and of the Environmental Agency (ANPA).

As far as public support for data is concerned, the picture is a mixed one. The survey shows that this is perceived as a medium expectation in all countries today (and a low requirement in the future in Italy, Germany and Sweden). However, our case-studies showed that support for a reliable source of data is perceived as a real need. This is especially true in Italy and – as could be expected – in SMEs. Our feeling is that larger collaboration between research institutes, governments and companies with regard to this specific issue could be very useful for the future of LCA, especially if it is to be used again in the future for marketing and eco-labelling purposes.

Policy's influence on the business' level of LCA-activities

The start of LCA within business is influenced by several factors. As mentioned above, clear patterns at the start of the institutionalisation process do not exist. Several factors are responsible for this. Policy pressure and environmental discussions are one factor amongst others. In Germany and Switzerland, we had some hints that environmental legislation (especially the German waste law) and governmental practices (especially public procurement) contribute to push LCA-activities within some companies (see subsection 5.4.4). However, the results of the survey show that environmental legislation is perceived as a driver of only medium importance. Especially in Sweden, environmental legislation is not an important driver for LCA[32].

Our general conclusion is that a new age of communication and co-operation needs to begin in the relationship between policy and business. Whilst a good culture of communication is present particularly in Sweden, in Germany and Italy, business and policy need to come closer together. New demands for the state to regulate environmental deterioration and problems need new forms of co-operation. However, there are some signs of such a process beginning. Examples are the Centre for Environmental Assessment of Product and Material Systems (CPM) in Sweden and the forthcoming Italian Association for LCA, both involving public and private institutions.

7.2 Interesting results

The objective of this section is to summarise the most interesting results, with respect to the process of integration of LCA in business, the role of LCA, the decisions arising from LCA and their implications for environmental policy:

[32] In the sense of command and control. On the other hand, there is a high level of communication between business and policy representatives.

Application trends of LCA in the four countries

As expected, the level of diffusion of LCA in the four countries is different. This result was derived from both the inventory and the survey. In absolute numbers, Germany showed the highest level of studies carried out. However, if this number is related to population and GDP, then Sweden presented as the country in which LCA is used most. Sweden also had the highest rate of returned questionnaires (41%, vs. 7.5% from Italy) and the highest absolute number of responding companies using LCA among the four countries (66 respondents). The "level" of use and application trends of LCA in the four countries can be briefly summarised as follows:

- *Germany:* Germany has, relatively speaking, a long-standing tradition in LCA. Since 1990, at least 16 LCA studies have been completed every year (with a peak of 28 in 1995). There is considerable interest in LCA, both in business and in political milieus. However, there are some misunderstandings and different view points: the confrontation between policy and business exists particularly concerning the role which LCA should play in the field of environmental policy; whereas business refuses such a role, environmental policy makers support it. As a consequence, a lot of debate and mistrust has arisen. Nevertheless, in general larger companies are very engaged in LCA, but with the main focus still being a rather retrospective one (LCA is usually used for learning and/or the confirmation of previous results). In general, German companies believe in the future increase of LCA activities, but often the underlying condition for this opinion is the combination of LCA with other instruments.
- *Italy:* The situation in Italy might be summed up as a good beginning. The knowledge and diffusion of LCA in this country has just recently begun. In many cases, international mother corporate companies triggered LCA studies. However, the situation seems to be evolving quickly. The numbers of studies and companies involved in LCA are both increasing. Most involved companies are developing internal LCA teams and know-how, and are optimistic about the future of LCA. There is a problem with SMEs, which have less motivation and resources to carry out LCA. However, Italy is in the process of founding an Italian Association for LCA, along the lines of the Japanese model. This association will include representatives of business, public administration, universities, research institutes, and NGOs. Its objective will be to tackle methodology and data issues and to reinforce communication and collaboration between companies, between business and policy, and between business and the research world. This is at the same time a way to support companies (particularly SMEs) and to create an innovative collaboration and communication climate between business and policy on this particular issue.
- *Sweden:* Sweden is the country among the four selected where LCA is most developed and used. This appears due to, amongst other things, a high level of environmental awareness in society, the proactive behaviour of companies, and a "cultural" tradition of collaboration between business and research. In addi-

tion, the relationship between business and public administration is very positive. On one hand, there is little intervention in the sense of command & control, on the other, there is a high level of communication. Sweden had the highest share of LCA carried out internally in firms, and the highest share of companies which expected an increased use of LCA in the future.
- *Switzerland:* Switzerland shows the second highest number of LCA studies per GDP unit (close to Sweden, and four times higher than Germany). There is certainly substantial experience of LCA in this country, where the public sector has been an important commissioning body. In a way similar to Sweden, the survey indicated a more pro-active tendency of companies to apply LCA to compare existing products with possible alternative ones. However, some saturation effects can also be observed This result is shown both in the case-studies and, more significantly, in the survey. 23% of companies responding declared that the use of LCA will not increase in the future.

Common patterns in the process of institutionalisation of LCA in all case-studies

The *institutionalisation theory* is able to explain successes and/or failures in the introduction of the LCA tool in business. We further observed that this process follows similar patterns in all countries. Of course, the marginal conditions differ from country to country, however the *process* can be described in the same way.

In this context, we stress the key *importance of actors* and subjective factors. We identified three crucial actors within the institutionalisation process, namely
- the initiator,
- the entrepreneur (or champion) and
- the set of champions.

Their relevance changes during the different process phases. The initiator is the central driver in the first phase. The importance of the entrepreneur (who might be, but is not necessarily, the same person as the initiator) is central in the objectification stage. His/her presence and influence are crucial factors for the overall success of the whole process. She/he is the one who develops a strategy for creating or emphasising a good justification for LCA and reaching a consensus within the company. Finally, in the sedimentation stage, when large consensus is needed, the set of champions involved in the diffusion of the tool has to expand to include different sectors of the company. The set of champions generally includes the entrepreneur, however this is not strictly necessary, since in the full sedimentation phase the tool "survives" even if the people change.

Another crucial point is the *organisational* context. In particular, for the realisation of a full institutionalisation process, good internal communication channels are very important.

For the future, we believe that process orientation will require two things:
- an integration of LCA into the chain of decision making in processes, and as a consequence,

- a simplification of LCA methods in order to adapt to the speed of decision making in process driven businesses.

We found these two tendencies in our case-studies.

A top-down or a bottom-up approach and the institutionalisation of LCA

The starting of LCA activities can occur both in a top-down or bottom-up approach. This is not significant for the further success of the institutionalisation process. Other factors are important, particularly during the second stage of objectification. Amongst these, there is a need to arrive at a consensus with or obtain the mandate of top-management at that stage.

Development of an internal know-how on LCA

External consultants can be a positive trigger for starting LCA activities in a company. However, following this, internal know-how has to be developed and LCA activities preferably carried out internally. Relying only on external consultants can be an obstacle and lead to failure of LCA activities. This result is confirmed by both the case-studies (all successful cases show established internal LCA teams and know-how, all companies relying only on external consultants faded out LCA activities) and the survey (e.g. 77% of Swedish responding companies carried out LCA internally).

Change in the role of LCA during the institutionalisation process: The importance of justification during the objectification stage

One big advantage of carrying out case-studies as an analysis method is that they allow the study of dynamic factors. One main outcome of this research is that the *role of LCA changes* over time during the different institutionalisation stages. At the beginning, the role played by LCA is usually for learning or confirmation and is not too important. However, during the objectification stage it becomes crucial. Indeed, whatever the exact role played in terms of application, LCA must provide a *good justification* for this. This is absolutely essential for the creation of consensus within the company. In all cases where a good justification was missing, LCA activities had an uncertain future or were stopped.

In particular, one major point is the ***perceived influence on the rest of the product life-cycle***. If companies perceive that they have no or only a limited influence on the rest of the product life-cycle, they see no reason for carrying out time-, money- and resource consuming LCA studies. In its turn, this perceived influence depends on several factors:

- size (larger companies influence easier smaller companies in the supply chain),
- sector (some sectors can be more appropriate for life-cycle management than others),
- position in the product life-chain, and

- structure of the company itself.

The present insufficient application potential of LCA for marketing purposes

In the past, especially in Germany, there have been great expectations concerning the use of LCA for marketing. However, one main observation from both the survey and the case-studies is that the *expectations* about the use of LCA for marketing could *not be satisfied*[33]. In many cases, marketing provided the first motivation for starting LCA, but companies quickly realised that LCA cannot be used for marketing. This is mainly connected to methodological issues, which often render results disputable. The other (related) problem is how to render in a very concise form a large and complex amount of data, information and assumptions[34].

This rather unexpected result had several *implications*. Where no learning process followed (see subsection 7.2.4), LCA activities faded out. Where a learning process occurred, either the use of LCA shifted towards more internal application, and/or it triggered the development of simplified/alternative tools.

In the future, we do not exclude LCA being used for marketing. In the long-term, companies may want to develop a tool able to provide economically tangible, short-term benefits. However, this requires the maximum of transparency and clear rules for simplification procedures, reporting, definition of system boundaries, all kinds of made assumptions, impact assessment methodologies, etc. Despite the great progress made within the framework of ISO standards, it is clear that a big effort is still needed in this direction. On the other hand, LCA can certainly be (and is already being) used to inform and influence suppliers and industrial clients. As shown by the survey, LCA is also used to provide generic information to customers and stakeholders (but not yet in Italy)[35].

LCA for learning

At present, perhaps the highest value of LCA is for learning. Both the survey and case-studies indicate that benefits of LCA are generally long-term. The clearest example of learning is the shift from external (marketing) motivations and unsatisfied expectations towards internal decision-making (the survey shows that the main applications of LCA are the identification of bottlenecks and research & de-

---

[33] In particular, there were big expectations, that LCA (eventually integrated with weighted aggregate environmental indicators) could be used to demonstrate that a product of the company A is environmentally better than the corrisponding product of the competitor company B.

It is the *perceived* degree of satisfaction about the fulfillment of expectations and the *perceived* possibility for change which do matter.

[34] This is particularly relevant in business-to-final consumer relations, less important with respect to business-to-business activities.

[35] In particular, Swedish companies use LCA as a basis for their environmental product declarations.

velopment activities). In our case-studies, we observed that where a learning process occurred, this could lead to:

- a change in internal organisation,
- the shift from the expected marketing use to internal applications (R,D&D, direct product development, choice of suppliers, etc.),
- spin-offs (energy and materials saving, bottleneck identification, cleaner and more efficient transportation systems, etc.),
- the development of simplified/alternative methodologies (i.e. ones focusing on those product life-cycle stages over which the company has most influence),
- a collaborative attitude (some companies may decide to not carry out full LCAs on their own, but to collaborate in sector- or country-specific studies; other might influence their suppliers and/or clients in this sense – i.e. Akzo Nobel).

This learning process depends very much on the actors and the flexibility of people with respect to unexpected results. All the ones who went through this process recognized and declared the strategic value of LCA, and the fact that it can be used for many different (expected and unexpected) purposes.

Implementation of LCA results

The translation of the results of the learning process into action can either directly or indirectly come from the learning cycle. As already mentioned, the benefits of LCA are rather long-term. This implies that there is a generally high acceptance of the integration of LCA results for *long-term* planning (strategic decisions, next cycle of investments, etc.). In the short term, LCA results are translated into action only if they lead to win-win situations (e.g. if they also imply cost-savings). In general, in the *short-term*, economic aspects always prevail.

Difficulties of SMEs carrying out LCA

Both the survey and the case-studies show that SMEs have particular difficulties in carrying out LCAs. This is likely to do with the lack of human and financial resources. One of the most important issues of LCA at present, i.e. the availability and quality of data, presents a particular difficulty for these companies. SMEs should be supported by policy and/or business association initiatives in starting and conducting LCA activities.

Implications for environmental policy measures

On the one hand, the results from our survey showed that the role of policy pressure is regarded as of minor importance for the adoption of LCA practice in companies. Less then 25% of the surveyed companies believed that LCA would be used in the near future for anticipating legislative threats. Major noted drivers for starting LCA were amongst others, cost saving opportunities, product related environmental problems and emerging green markets. The results of the case studies we carried out confirmed the above-mentioned outcomes. Compared with the other listed motivations, regulatory pressure is not the most important factor for stimulating LCA-activities.

On the other hand, if required, LCA is generally considered to be a suitable tool for responding to external pressure from regulation, and for anticipating legislation. We observed that several companies introduced LCA because of company-external discussions and pressures (existing, expected or perceived).

Morevoer, policy can support LCA activities in either direct or more indirect ways. Examples of direct support are green public purchasing, which has been an important triggering factor in Switzerland, and the financing of LCA studies. However, probably the most important support is indirect: policy can be very effective in developing public life-cycle inventory data-bases, supporting eco-labels, supporting ISO activities, collaborating with business in solving methodological issues (i.e. related to impact assessment). This kind of help is generally well accepted by companies. In contrast, companies strongly mistrust the possible use of LCA as the foundation of command & control policy measures.

Our general conclusion about the relationship between policy and business is that a new age of communication and co-operation needs to begin. While a good culture of communication is present particularly in Sweden, in Germany and Italy, business and policy need come closer together. New demands of the states to regulate environmental deterioration and problems need new forms of co-operation. Examples of this are the Centre for Environmental Assessment of Product and Material Systems (CPM) in Sweden and the recently founded Italian ssociation for LCA, where both public and private institutions are involved.

Need for collaboration

Our last finding is that more general collaboration is vigorously needed for a successful and effective diffusion of LCA. Collaboration is a must within a single company: different departments, actors and champions must communicate with each other, to support each other and try to learn together from LCA. A reasonable result might be a successful learning process leading to a decision to continue with LCA-activities. Besides that, as a second type of collaboration is required in the supply-chain, where upstream and also downstream collaboration is necessary to exchange environmental information and also to agree on measures to improve the eco-profiles of products. A third type of very valuable collaboration is the exchange of ideas, opinions and experiences among researchers and practitioners (this type of collaboration already exists in Sweden and Switzerland, but is still not common practice in Germany; it is starting in Italy, again in connection with the Italian Association for LCA).

7.3 Recommendations

Our research outcomes lead us to the formulation of a number of recommendations. These refer first to the policy and governmental level (subsection 7.3.1). The penultimate subsection 7.3.2 introduces some research questions for further R&D activities in this area. The last subsection 7.3.3 presents some recommendations to business for dealing with LCA.

7.3.1 Policy recommendations[36]

The recommendations presented in this subsection are addressed to the people and functions responsible for product policy at the European and national administrative levels. Before presenting some detailed points, we want to make some overall recommendations:

1. We believe that the use of LCA should not be imposed on companies by policy. The best way to diffuse the application of LCA is through competition. The proactive and innovative behaviour of companies should not be limited, but be stimulated. In other words, LCA should become a tool for companies to obtain comparative advantages (either short-term and/or long-term) on the market. How companies acquire this advantage is not a policy affair.
2. The responsibility of policy is to create and guarantee a fair level of competition. In particular, this refers to
   - improving and guaranteeing the availability and quality of data, in particular with respect to public or general services (e.g. energy, transportation, waste, main group of materials),
   - supporting SMEs without sifficient financial and human resources to carry out LCA,
   - monitoring and guaranteeing a high-level standard of LCA data reporting,
   - collaborating and communicating with business to solve LCA-related problems,
   - supporting ISO activities which are very well accepted by business.
3. We do not agree with the business attitude that LCA should not be used for legislation and environmental policy. We do not rule out the possibility that some important policies (e.g. energy policy, transport policy and waste policy) which of course affect business might also be strengthened by the use of LCA.
4. Our research leads us to the key question of policy and business: The role of LCA is unclear and there exist many misunderstandings about it. We think that both policy and business have to move from a position of confrontation towards new forms of co-operation and communication in order to solve together many of the challenges listed below.

---

[36] See also Smith et al. (1998) who developed some proposals for policy options; they have been divided into two main categories "Reduce barriers" and "Encourage fair play".

We have subdivided our specific proposals into five clusters of measures which are also presented in Figure 7.4:

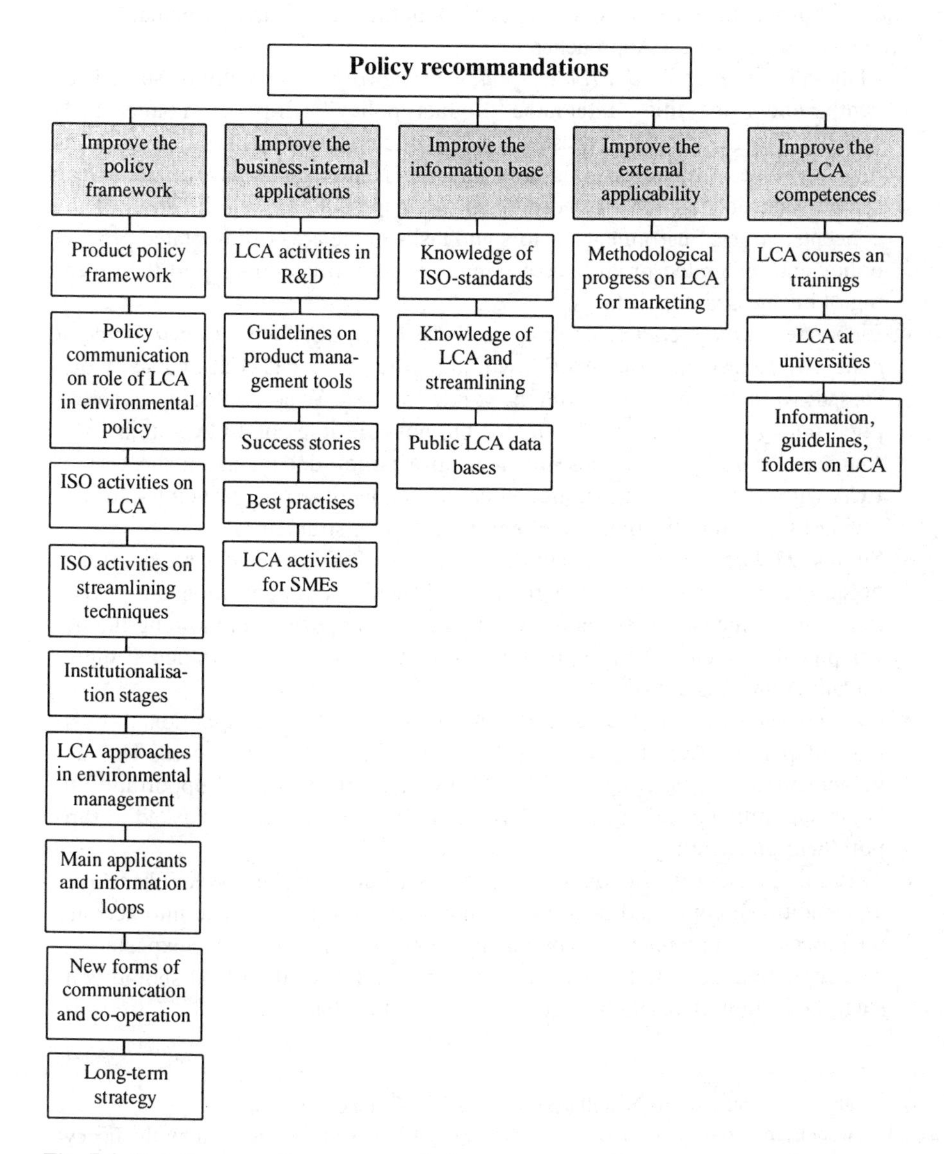

Fig. 7.4. Overview on policy recommendations

Improve the policy framework

This category encompasses measures which are intended to elaborate, clarify and/or improve the existing policy framework in the area of environmental policy, products and Life Cycle Approaches:

- Elaborate an *Integrated Product policy framework*: This activity started recently under the title "Integrated product policy" (Ernst & Young/SPRU 1998)[37]. The EC intends to present a policy document on this topic; some of its member states (especially Denmark and the Netherlands) have already presented such documents[38]. According to this approach, any clarification of the concepts, targets, instruments, actors and tasks in the area of integrated product policy and integrated product management is appropriate to support the greening of business and their tools.
- Elaborate a *policy communication* on the *role* of LCA within *environmental policy*: Currently, the role of LCA within public policies is controversially discussed; we have identified different policy "styles". What is missing is an explicit policy document which stresses the possible roles of LCA within environmental policy. We propose to develop a policy document on the role of LCA within environmental policy which also contains some principles for applying LCA within the plethora of environmental instruments.
- Support *ISO-activities*: LCA activities within the ISO framework are necessary because they create an unambiguous world-wide communication basis. It is important to increase their public acceptance by supporting and joining the development as such and by supporting the participation of governmental representatives and NGOs.
- Push *ISO-activities* on the *streamlining* of LCA: As already mentioned, business adapts or streamlines Life Cycle Approaches to their purposes. This development should be considered by the ISO and offers a good opportunity to begin supporting standardisation activities in this area[39]. Policy is asked to support them practically.
- Take into account *institutionalisation stages* while preparing *policy measures*: The traditional command & control policy was not forced to take into account the process of implementation by business practices. However, the expected increasing stimulation and support of "soft" instruments (like LCA) should consider the institutionalisation stages and try to reduce barriers.

---

[37] See also Oosterhuis/Rubik/Scholl (1996) and Rubik/Teichert (1997).

[38] In Sweden, important contributions in the area of IPP have been prepared by the Ecocycle Commission and are represented in the bill of producer responsibility.

[39] As already mentioned, it is important to notice, that ISO is seriously addressing the question of simplification rules. In particular, recognising that the issues of simplification might be sector-specific, in a recent workshop of CEN it was proposed to establish working groups on LCA focusing on particular industry sectors. Part of their work should be also to establish simplification rules.

- Strengthen *Life Cycle Approaches* in *environmental management*: The EMAS-scheme possesses some modest hints towards products and LCA. We propose to enlarge these and to mention LCA within the EMAS-directive as a tool which is able to support environmental management systems.
- Take a *look* at the *main applicants* of LCA and *close information loops*: We identified a certain mismatch between expectations of policy makers concerning the application of LCA by business and the real applications of LCA by business. These open information loops should be closed and used for more precise and target-oriented governmental measures in the area of LCA.
- Establish *new forms* of *communication* and *co-operation*: The existing ways of co-operation between policy, stakeholders and business are not sufficient to establish and support LCA-activities. We propose the establishment of new forms of co-operation based on the example of the recently founded Italian LCA association[40] or the Swedish CPM organisation, established in 1996, where business, universities and policy makers are represented.
- Have a *long-term strategy*: Policy will be disappointed if it expects immediate environmental gains from business' application of LCA. On the contrary, LCA needs a long-term strategy, both in business and policy.

Improve the business-internal applications

Policy is able to support business on its way to greening products; some policy initiatives and measures which we think are suitable for this purpose are listed below:

- Consider *LCA-activities* in *R&D policy*: Business' R&D activities for product innovation might be supported by public funding. This funding is a good opportunity to introduce/enlarge environmental assessment tools (like LCA) into business. R&D funds should support these processes by offering funds for technical help; however, the approach should be a liberal (an "offer") instead of a mandatory one (a "requirement").
- Elaborate *guidelines on product management tools*: People involved in product design and development process need adequate environmental assessment tools and knowledge of their strengths, characteristics and application. A lot of tools exist which might fulfil this objective; one among them is LCA. Therefore (generic) guidelines should be elaborated which support people involved and demonstrate best practices on the application of different product management tools. Being aware that there exist hundreds of general guidelines, we propose to develop specific guidelines, in this case oriented to specific products.
- Publish *success-stories*: As mentioned, LCA-activities need justification for successful implementation. We think that good examples exist, but they are not really known of and communicated. Reported types of success stories might be

[40] The Association was officially established in January 1999. However, unofficial working groups have already been active since the beginning of 1998.

based on the institutionalisation process (What happened, how? Why was the introduction of LCA successful or not?), economic profits (win-win situations), reductions of environmental burdens etc.
- Report *best practices* and *cost-savings opportunities* through the application of LCA: similar to the publication of success stories, best practices should be reported encompassing best practices of LCA-applications within product design, product development, procurement and marketing.
- *Support LCA*-activities for *SMEs*: SMEs are still not very active in the field of LCA. We propose to initiate a specific programme for SMEs which provide subsidies and training for SMEs staringt LCA-activities.

Improve the information basis

An adequate policy framework and the company's intention to strengthen life cycle considerations is a necessary but insufficient condition for the successful introduction and application of LCA. One major further requirement is an adequate information basis. Indeed, both the survey and the case-studies showed that the availability and quality of data is still one of the major issues of LCA at present. Another major issue is how the data and results have to be treated and presented. We propose the following measures:
- Increase the *public knowledge* of *ISO-standards*: ISO-standards are known, but the knowledge of ISO-standards on environmentally oriented product management is insufficient. The dissemination of ISO-standards of the 14040-series has to increase. Environmental agencies are asked to initiate target group oriented campaigns.
- Increase the *knowledge* of the tool *LCA* and *streamlining techniques*: The complex tool LCA is not very well known and understood by companies. In addition to campaigns on ISO-standards, environmental agencies should inform companies on LCA in general. Moreover, there is a clear need for agreed and common simplification procedures within business. Diffusing information abour streamlining techniques would be a great help to companies.
  An important prior target group are SMEs.
- Expand and improve *public LCA-data bases*: The availability of data bases is restricted. Only for some commodities, materials and product groups, have data been published. Their existence should become more known and easier to access; in addition to this, other important data should be elaborated on both European and national levels.

Improve the external applicability

The greening of business, its activities and its products should also be communicable for business-external purposes. The improvement of the external applicability of LCA should be supported by the following measures:

- Support *methodological progress* on the *application* of LCA for *marketing*: Companies might be interested in also applying LCA for marketing and/or external communication. They need in the short run some additional guidelines and rules; therefore, an ISO-standard should be elaborated with this in mind. In the long run, methodological progress should also be directed towards these application possibilities. An example of ongoing work in this area is the task-group working on the draft for the "Type III Ecolabelling" ISO standardisation.

Improve LCA-competencies

LCA needs to be put into practice. Actors and people are necessary for this. They need adequate competencies. Policy is able to support this by the following measures:

- Support LCA-*courses* and *training*: Most potentially relevant employees of business neither know about LCA or are not familiar with the technique. We propose that funding for specific training will be offered.
- Teach LCA at *universities*: The curriculum of universities should include the teaching of environmental assessment tools like LCA. Hereby, the number of well-educated people might increase; as a consequence they might contribute to a more intensive dissemination of LCA-techniques.
- Present *information, guidelines* and *folders* on LCA: The self-teaching of business' staff should be supported by guidelines and folders which inform on LCA.

7.3.2 Research recommendations

The R&D society is also asked to pursue additional R&D activities. We have structured these into three clusters and propose the following important R&D activities:

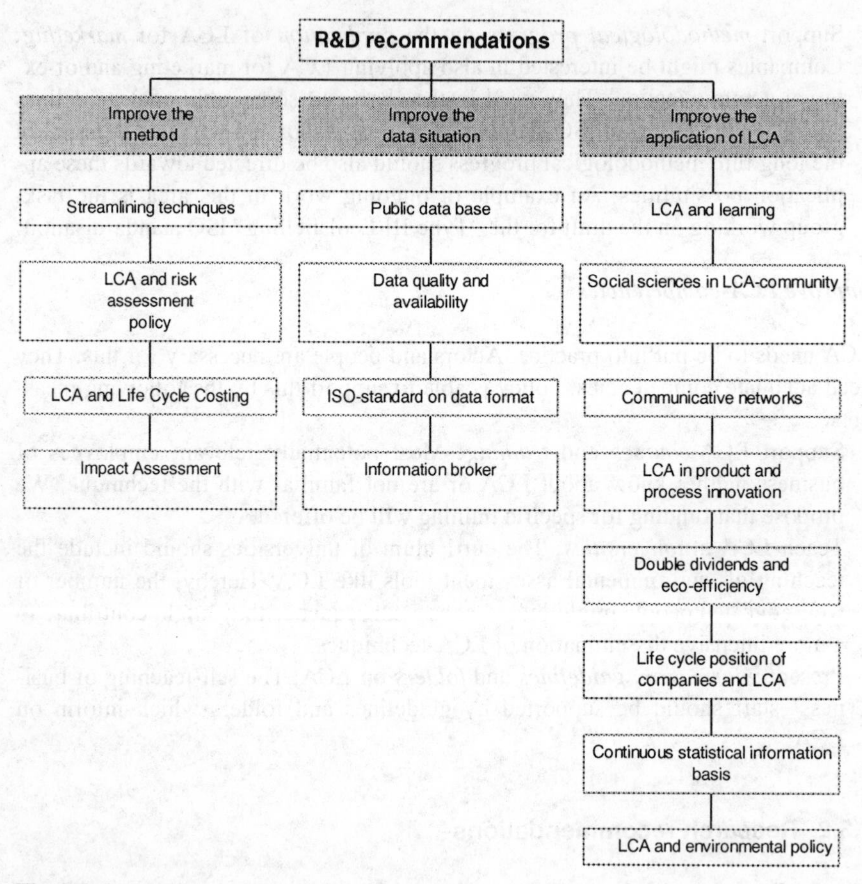

Fig. 7.5. Overview on research recommendations

Research on improving the method

- *Streamlining techniques*: A clear message of our research is that Life Cycle Approaches should be adopted for business targets. But are there any clear recommendations which might reduce the time and money invested in these adoption/streamlining processes? As already mentioned, streamlining and/or simplification guidelines migth be sector-specific.
- *LCA and risk assessment*: Not too much work on this issue has been done. We propose a study directed at the insurance sector (whether and how they apply LCA for assessing environmental liabilities and environmental risks - for example of climate change).
- *Life Cycle Costing* and *LCA*: One of the expectations about LCA is that it might be a helpful basis for internalising environmental externalities within business

activities, thus increasing its role as a decision support tool for business. However, further R&D is needed to connect LCC with LCA. For example, one criticism of LCA's Impact Assessment is that in fact it does not assess the real environmental damages nor does it encompass information about the response of the local environment. Further scientific work is needed to distinguish potential and real impacts. More studies are needed about the methods to determine life-cycle externalities (i.e. risk assessment, willingness to pay, Environmental Priority System (EPS), taxation, etc.) and to connect them with LCA data.

- *Life Cycle Impact Assessment*: We propose more research on the use of multi-criteria analysis.

Research to improve the data quantity and quality

- *Public data base*: One main barrier for LCA-studies is missing/insufficient data along the life-cycle. Some collaborative studies (e.g. on behalf of APME, of BUWAL, of UBA etc) have tried to introduce some inventory data of important substances/products/materials. Also the activities of SPOLD and CPM are important. However, a systematic overview on existing LCA-data is missing and there is also a lack of specific data on some important upstream processes. We propose the enlargement of data-generating activities, accompanied by clear priorities on the most important products/materials. Amongst other things, the development of routines for cost-efficient data generation is taking place within the industrial network of CPM.

  A specific emphasis has to be focussed on the needs of SMEs.
- *Data quality* and *availability*: Standards for minimum qualities of data should be elaborated and fixed. Provided data should be as much disaggregated and detailed as possible. In addition to this, new methods for filling in missing data (e.g. fuzzy logic, energy and mass-balances) should be developed.
- ISO-*standard* on *data-format* within the 14040-series: Some years ago, SPOLD published a draft version of a data-base format (Singhofen 1996); also CPM published a report about the SPINE data base format (Stehen et al. 1995). The work with the ISO 14041 – LCA dataformat is ongoing. The working group (ISO TC207/SC5/WG2) use the SPOLD and SPINE formats as a foundation for the standardisation. These activities should be accompagnied by R&D to minimise transaction costs and quality loss.
- Information *broker/adviser*: A lot of environmental data based on LCA is already available. However, the transparency and overview of this data market is incomplete. An information broker might reduce information barriers[41].

[41] Data trading is already established between the CPM corporations. A public trading place for LCA data has been set up by CPM in relation to their LCA data base and education in LCA data documentation is given to companies and branch institutes. Experience from this data trade could be useful for further trading expansion.

Research on the applications of LCA

- *LCA* and *learning*: One of the major results of this study is the highlighting of the importance of LCA for learning. Our integration of LCA within the framework of institutionalisation theory is one first step in this direction. More work is needed to explore how companies can develop their capability (both at a technical and an organisational level) to fully exploit the learning potential provided by LCA.
- Strengthening *social sciences* in research and discussion on the method of LCA: The predominant focus of LCA-research has been oriented towards the improvement of the method as such. However, the social embedding of the technique has often been "forgotten" or was not important. But the personnel and social context determines the success or failure of the concept. Therefore, social science is asked to consider the institutionalisation processes of an innovation like LCA (in other words to underline the importance of organisational innovation).
- *Communicative networks*: One of the most instructive results is the importance of the actors. The analysis of the networks of relationships both within a company and with the external world still remains an important R&D task.
- Impacts of *LCA* in product and process *innovation*: Current knowledge on the relationship between LCA-activities and product/process innovation is modest. Some first hints at a positive relationship exist, but they are not systematised and related to barriers/opportunities. An exhaustive empirical study is necessary in this area.
- Impacts of *LCA* on double dividends and *eco-efficiency* strategies: It is still not clear whether LCA contributes to the realisation of double dividends (namely economic and environmental ones) and whether it increases eco-efficiency. This question might be the topic of another research project. Specific emphasis should be focussed on measures to increase opportunities and to reduce barriers.

  Another important challenge is the assessment of environmental profiles with regard to the shift from selling products to delivering services. Is this a general cultural issue or are there other problems? Is it sector-dependent?
- The *life cycle position of companies* and *LCA-activities*: As described by Smith et al. (1998) and by us, the position along the life cycle of products might influence the level and type of LCA-activities. However, the findings generated so far are ambiguous.
- Enlarge *information basis*: The application of LCA by business is relatively well known in the case of some countries. But some important countries (e.g. France, United Kingdom) are missing and the information gained might be out of date. We propose a wider report on industrial LCA activities as part of surveys.

- *LCA and policy*: The role of LCA is not clear as a basis for environmental policy actions. Arguments in favour and against are often used, but the conceptual and empirical basis is insufficient. Therefore, this too is a task for R&D.

7.3.3 Business recommendations

Beside policy and research, we have some precise recommendations for business as to how to proceed with LCA and how to start/institutionalise LCA-activities, namely:

How to consider LCAs

- There are several instruments, which might be used for an environmental assessment of products and services. Choose the tool(s), which might offer the most opportunities for you. There are several Life Cycle Approaches which encompass a broad range from Life Cycle Thinking (LCT) to Life Cycle Assessment (LCA).
- LCA is one prominent tool, but very likely it has to be adopted ("streamlined") to your purposes.
- The environmental situation is always complex. LCA confronts you with a consideration of the life cycle impacts of your products across environmental media. It offers new ways of looking at products and services!
- LCA contributes to a better understanding of environmental complexity; it designs situations, but it does not solve them!
- LCA might play different roles along the product development chain. Be aware of them!

How to introduce and develop LCA activities within the company

- Starting LCA-activities and bringing them into your business is a step by step process. Start with learning; later your focus will move to doing! Learning might mean that you concentrate on getting knowledge of the method; restrict yourselves and start by assessing "simple" products instead of complex ones!
- It is not important whether the start of your activities follows a bottom-up or top-down approach. Just start!
- Learning means a general attitude of being open to new insights, findings and development. Be curious about any (also unexpected) results!
- The objective of your LCA-activities should be to institutionalise them within the company, provided that this makes sense.
- Therefore, you need people and actors actively pushing the LCA development within your company! You need an entrepreneur who has a high personal commitment who is able to influence the company and who is persuasive.

- The central task of the entrepreneur is to organise the LCA process, to look for support within the company, to create links, to communicate and to spread out LCA within the company; hereby creating a large set of champions.
- In addition, an appropriate justification for LCA-activities is necessary.
- Develop company-internal know-how which is formally implemented into business' activities. Develop capabilities for carrying out LCAs internally (relying only on external consultants will not work)!
- If you have learned enough, some sort of formal decision on the future on LCA within the company has to be taken. This means that a mandate of top-management should be organised.
- Do not expect LCA to be *immediately* applied to marketing. Your prior orientation should be a business-internal application. However, LCA can be used for supply-chain management and to influence industrial clients. Transparent and coherent external communication is also allowed.
- Your product innovation strategy should be a long-term one. Do not forcely expect LCA to deliver results to you which might immediately be applied (although some cases might exist).

Other recommendations

- Participate in and support new forms of collaboration between business and business, between business and research and between business and public administration.

References

Ackhoff, R.L. (1979): The Future of Operational Research is Past. In: Journal of Operational Research Society, No. 2, pp. 93-104

Ahbe, S. / Braunschweig, A. / Müller-Wenk, R. (1990): Methodik für Ökobilanzen auf der Basis ökologischer Optimierung. Bern: BUWAL - Schriftenreihe Umwelt No. 133

Allen, D.T. et al. (1997): Public Policy Applications of Life-Cycle Assessment. Brussels: SETAC-Technical Publications Series

Andersson, T. / Wolff, R. (1996): Ecology as a Challenge for Management Research. In: Scandinavian Journal of Management, No. 3, pp. 223-231

Ankele, K. / Steinfeldt, M. (1995): Ökobilanz von Verpackungssystemen. Berlin: study of the Institut für ökologische Wirtschaftsforschung on behalf of Weleda AG

Argyris, C. / Schön, D.A. (1976): Theory in Practice: Increasing Professional Effectiveness. San Francisco: Jossey Bass

Arnold, F.S. (1993): Life cycle doesn't work. In: The Environmental Forum, No. 5, pp. 19-23

Årsredovisning 1995/96 (1996): Perstorp, Corporate Communications, Klippan Tryckeri

Ayres, R.U. (1994): Life Cycle Analysis - A Critique. Fontainebleau/F: Working paper No. 94/33/EPS - Center for the Management of Environmental Resources (CMER)

Bakker, C. (1995): Environmental Information for Industrial Designers. Delft/NL: PhD thesis, Technical University of Delft

Baumann, H. (1996): LCA Use in Swedish Industry. In: International Journal of LCA, No. 3, pp. 122-126

Baumann, H. (1998): Life Cycle Assessment and Decision Making - theories and practises. Göteborg: Chalmers University of Technology AFR Report 183

BDI <Bundesverband der Deutschen Industrie> (1996): BDI-Position für den geplanten UBA/BMU-Workshop "Bewertung von Ökobilanzen". Köln: BDI Position Paper

BDI <Bundesverband der Deutschen Industrie> (1998): Innovation in the Market Economy. The Political Role of Life Cycle Assessments. Köln/D: BDI Position Paper

Becker, T. (1997): Informationsinstrument für eine innovations- und nicht interventionsorientierte Politik. Chance für Kooperation. In: Ökologisches Wirtschaften, No. 6, p. 24

Becker, T. / Marsmann, M. (1998): Position der Industrie zur "Bewertung in Ökobilanzen", in: UTECH (1998): Ökobilanzen VI. Berlin: FGU, pp. 55-73

Beckman, T. / Baumann, H. (1998): The use of LCA in Business decision-making. Göteborg (S): TMP-Report 1998:2

Belz, F. / Strannegárd, L. (1997): International business environmental barometer. Oslo: Cappelen Akademisk Forlag

Bengtsson, M. (1998): Värderingsmetoder i LCA. Göteborg: Technical Environmental Planning

Berkhout, F. (1996): Life cycle assessment and innovation in large firms. In: Business Strategy and the Environment, No. 5, pp. 145-155

BMU <Bundesministerium für Umwelt; Naturschutz und Reaktorsicherheit> (ed.) (1998a): Nachhaltige Entwicklung in Deutschland. Entwurf eines umweltpolitischen Schwerpunktprogramms. Bonn

BMU <Bundesministerium für Umwelt; Naturschutz und Reaktorsicherheit> (ed.) (1998b): Umweltgesetzbuch (UGB-KomE). Entwurf der Unabhängigen Sachverständigenkommission zum Umweltgesetzbuch beim Bundesministerium für Umwelt; Naturschutz und Reaktorsicherheit. Berlin: Duncker & Humblot

BMU <Bundesministerium für Umwelt; Naturschutz und Reaktorsicherheit> (1999): Informelles EU-Umweltministertreffen in Weimar vom 7. Bis 9. Mai 1999. In. Umwelt, No. 6

Boustead, I. (1996): LCA - How it came about. In: The International Journal of LCA. No. 3, pp. 147-150

Bragd, A. / Wolff, R. (1996): Differentation in Environmental Marketing. Gothenborg/S: Gothenburg Research Institut GRI Working Paper

Brunsson, N. (1994): The Irrational Organisation. New York/London: Wiley

Brunsson, N. / Olsen, J.P. (1993): The Reforming Organization. London: Routledge

Bültmann, A. / Rubik, F. (1999): Selected case-studies on the application of LCA within German companies. Berlin: Publication series of the Institut für ökologische Wirtschaftsforschung [forthcoming]

BUS <Bundesamt für Umweltschutz> (1984): Ökobilanzen von Packstoffen. Bern/CH: Schriftenreihe Umweltschutz No. 24

BUWAL (1992): Vergleichende ökologische Bewertung von Anstrichstoffen. Bern/CH: BUWAL-Schriftenreihe No. 186

Commission des Communautés Européennes (1992): Verzeichnis der Verbände in der Europäischen Gemeinschaft - 5th edition 1992. Brussels: Editions Delta

Cowell, S.J. (1997): Life Cycle Assessment: From Tool to Process. Presentation at SETAC-Europe Conference. Amsterdam, April 6-10, 1997

Curran, M.A. (1996): Environmental Life-Cycle Assessment. New York: McGraw-Hill

Curran, M.A. (1997): Life-Cycle Based Government Policies. In: The International Journal of Life Cycle Assessment, No. 1, pp. 39-43

Czarniawska, B. / Wolff, R. (1986): How we decide and how we act: On the assumptions of Viking Organization Theory. In: Wolff, Rolf (ed.): Organizing industrial development. Berlin: Walter de Gruyter, pp. 281-303

De Smet, B. et al. (1996): Life Cycle Assessment and conceptually related programmes. Brussels: Draft Paper - SETAC-working group on Conceptually Related Programmes

DeSimone, L.D. / Popoff, F. (1997): Eco-efficiency: the business link to sustainable development. Cambridge/London: MIT Press

Dosi, G. (1988): The nature of the innovation process. In: Dosi, G. et al. (eds.): Technical Change and Economic Theory. London: Pinter, pp. 221-238

EC <European Commission> (1998a): 2nd amended proposal for a European Parliament and Council decision conserning the 5th framework programme - Draft working document COM(98) 8 (March 1998). Brussels: European Commission

EC <European Commission> (1998b): Proposals for Council Decisions concerning the specific programmes implementing the fifth framework programme - Draft working document COM(98) 305 (July 1998). Brussels: European Commission

Ecobilan (1994): The eco labelling criteria based on the life cycle inventory of eleven indoors decorative paints. Paris: Study of ecobilan

Ehrenfeld, J.R. (1997): The Importance of LCAs - Warts and All. In: Journal of Industrial Ecology, No. 2, pp. 41-49

Elsener, B. / von Flotow, P. / Temme, C. (1998): Auswertung weiterer empirischer Studien zur praktischen Umsetzung der EG-Umwelt-Audit-Verordnung. Oestrich-Winkel: manuscript of Institute for Environmental Management and Business Administration [forthcoming]

Enquete-Kommission "Schutz des Menschen und der Umwelt" des Deutschen Bundestages (ed.) (1993): Verantwortung für die Zukunft. Wege zum nachhaltigen Umgang mit Stoff- und Materialströmen. Bonn: Economica

Erasmus, H. (1993): EC Legislation: Legislation, Priority Waste Streams, Transport Regulations Quality Assurance, Certification etc.. The Hague/NL: Presentation to the European Conference on Hospital Waste Management 8-9 December at Utrecht

Ericsson (1996): Ericsson Corporate Environmental Report. Stockholm: own publication

Ernst & Young / SPRU (1998): Integrated Product Policy. London: Study on behalf of European Commission DG XI

eurostat (1996): Statistische Systematik der Wirtschaftszweige in der Europäischen Gemeinschaft. NACE Rev. 1. Luxemburg: Amt für amtliche Veröffentlichungen der europäischen Gemeinschaft

FEU <Forschungsgruppe Evaluation Umweltaudit> (1998): Umweltmanagement in der Praxis. Berlin: UBA-Texte 52/98

Fichter, K. (1998): Umweltkommunikation und Wettbewerbsfähigkeit. Marburg/D: Metropolis

Fink, P. (1997): The Roots of LCA in Switzerland. Continous Learnning by Doing. In: The International Journal of LCA, No. 3, pp. 131-134

FTU/VITO/IÖW (1995a): Eco-balances: the uses and limitations of a tool for consultation and for decision-support. Brussels: Report to the King Baudouin Foundation

FTU/VITO/IÖW (1995b): L'écobilan pourquoi et comment: un approche pratique. Brussels. Roi Baudouin Foundation

Fussler, C. (n.y.): Life-Cycle Assessment - A new Business Planning Tool? Brussels: publication of SPOLD

Gabel, H.L. et al. (1996): Life-Cycle analysis and policy options: The case of pulp and paper industry. Business strategies and the Environment. John Wiley & Sons

Galbraith, J. R. (1973): Designing Complex Organisations. Reading, Mass.: Addison-Wesley

Gege, M. (1997): Kosten senken durch Umweltmanagement. 1000 Erfolgsbeispiele aus 100 Unternehmen. München: Verlag Franz Vahlen

Giegrich, J. et al. (1995): Bilanzbewertung in produktbezogenen Ökobilanzen. Evaluation von Bewertungsmethoden, Perspektiven. In: Umweltbundesamt (ed.): Methodik der produktbezogenen Ökobilanzen - Wirkungsbilanz und Bewertung. Berlin: UBA - Texte 23/95

Grießhammer, R. / Bunke, D. / Gensch, C.-O. (1997): Produktlinienanalyse Waschen und Waschmittel. Berlin: UBA-Texte 1/97

Grotz, S. / Rubik, F. (1997): Bibliographie Produktbilanzen. Berlin: Publication series of the Institut für ökologische Wirtschaftsforschung No. 92/97

Grotz, S. / Scholl, G. (1996): Application of LCA in German Industry. Results of a Survey. In: The International Journal of Life Cycle Assessment, No. 4, pp. 226-230

Hales, C.P. (1986): What do Managers do? A critical review of the evidence. In: Journal of Management Studies, No. 1, pp. 88-115

Hedberg, B. / Wolff, R. (1999): Organizing, Learning, and Strategizing. From Construction to Discovery. To be published in "Handbook on Organizational Learning". Oxford: Oxford University Press
Henkel (n.y.): Umwelterklärung Henkel KGaA. Düsseldorf-Holthausen: own publication
Hirschhorn, L. / Gilmore, T. (1992): The New Boundaries of the 'Boundaryless' Company. In: Harvard Business Review, No. 3, pp. 104-115
Hitt, M.A. / Tyler, R.B. (1991): Strategic decision models: Integrating different perspectives. In: Strategic Management Journal, pp. 327-351
ISO (1996): Environmental management - Life cycle assessment - Principles and framework - ISO 14040. Paris/F: International Standardisation Organization [ISO]
ISO (1997): Environmental management - Life cycle assessment – Goal and scope definition and inventory analysis - ISO 14041. Paris/F: International Standardisation Organization [ISO]
Jensen, A.A. et al. (1997): Life Cycle Assessment. A Guide to Approaches, experiences and information sources. Copenhagen: European Environment Agency – Environmental Issues Series No. 6
Jönsson, S.A. (1995): Goda utsikter: svensk management i perspektiv. Stockholm: Nerenius & Santérus
Jönsson, S.A. (1996): Perspectives on Scandinavian Management. Göteborg/S: BAS
Klöpffer, W. / Renner, I. (1995): Methodik der Wirkungsbilanz im Rahmen von Produkt-Ökobilanzen unter Berücksichtigung von nicht oder nur schwer quantifizierbarer Umwelt-Kategorien. In: Umweltbundesamt (ed.): Methodik der produktbezogenen Ökobilanzen - Wirkungsbilanz und Bewertung. Berlin: UBA - Texte 23/95
Klüppel, H.-J. (1993): Ökobilanzen von Waschmitteln - Perspektiven und Grenzen. Contribution to the annual conference of SEPAWA, Bad Dürkheim 14-15 October 1993
Klüppel, H.-J. et al. (1995): Life-cycle assessment of palm alcohol sulfate. In: INFORM, No. 6, pp. 647-657
Knoepfel, P. (1997): Switzerland. In: Jänicke, M. / Weidner, H. (eds.): National Environmental Policies. A Comparative Study of Capacity-Building. Berlin/Heidelberg/New York: Springer Verlag, pp. 175-197
Kotler, P. (1991): Marketing Management. Analysis, Planning, Implementation and Control. New York et al: Englewood Cliffs
Lindfors, L-G. et al. (1995): Nordic Guidelines on Life-Cycle Assessment. Copenhagen: Nordic Council of Ministers - Report Nord 1995:20
March, J.G. (1981): Decisions in Organizations and Theories of Choice. In: Van de Veen, Andrew / Joyce, William (eds.): Assessing Organizational Design and Performance. New York, NY: Wiley Interscience
March, J.G. (1986): Theories of making choice and making decision. In: Wolff, Rolf (ed.): Organizing industrial development. Berlin: Walter de Gruyter
March, J.G. / Simon, H.A. (1958): Organizations. New York: Wiley
Meier, U. / von Däniken, A. (1998): The use of LCA in business decision-making processes. Step 3.1: Swiss survey. Zürich/CH: manuscript of ökoscience Beratung AG [not yet published]
Ministry of Environment (Sweden) (1997): From Environmental Protection to Sustainable Development. Stockholm/S: own publication
Mintzberg, H. (1973): The Nature of Managerial Work. New York: Harper and Row
Mirulla, R. (1998): The use of LCA in business decision-making processes. Step 3.1: Italian survey. Milano: manuscript of Ambiente Italia

Neitzel, H. (1997): Die Bedeutung von Produkt-Ökobilanzen in der staatlichen Umweltpolitik. Instrument der Entscheidungsfindung. In: Ökologisches Wirtschaften, No. 6, pp. 22-23

Neumann, U. / Franze, H.A. (1996): BMW Status on LCA in the Automotive Industry: management report for decision finders in Policy and Industry. Munich: internal document.

Nonaka, I. / Takeuchi, H. (1995): The Knowledge-Creating Company. New York, NY: Oxford University Press

Oberbacher, B. et al. (1974): Abbaubare Kunststoffe und Müllprobleme. Berlin: Erich-Schmidt-Verlag - Beiträge zur Umweltgestaltung - No. A 23

Oberbacher, B. et al. (1996): LCA - How it Came About. In: The International Journal of Life Cycle Assessment, No. 2, pp. 62-65

OECD (1995): The Life Cycle Approach: An Overview of Product/Process Analysis. Paris: OECD own publication

OECD (1997): Eco-labelling: Actual effects of selected programmes. Paris: OECD-document No OCDE/GD (97)105

OECD (1998a): Eco-efficiency. Paris: OECD

OECD (1998b): Review of the Development of International Environmental Management Systems. Paris: OECD - ENV/EPOC/PPC(98) 6

Oosterhuis, F. / Rubik, F. / Scholl, G. (1996): Product Policy in Europe: New Environmental Perspectives. Dordrecht/Boston/London: Kluwer Academic Publishers

Pfeiffer, W. (1983): Strategisch orientiertes Forschungs- und Entwicklungsmanagement - Probleme und Lösungsansätze aus Sicht der Wissenschaft. In: Blohm, H. / Danert, G. (eds.): Forschungs- und Entwicklungsmanagement. Stuttgart: J.B.Metzeler / C.E.Poeschel, pp. 124-133

RDC [Recherche Developpement and Consulting] / Coppers & Lybrand (1997): Ecobalances for policy making in the domain of packaging and packaging waste. Brussels: Study on behalf of DGXI

Rentsch, C. (1994): Environmentally oriented Product Policy in Switzerland. Bern: Paper presented at the OECD/PPCGH Special session on Life cycle Management and Sustainable Product Policies November 2, 1994

Rubik, F. (1998a): Application Patterns of Life Cycle Assessment within German Companies. Results and conclusions of a Survey. Berlin: IÖW-Schriftenreihe No. 129/98

Rubik, F. (1998b): Step by Step. Ein Vorschlag zur Stärkung des Produktbezugs von EMAS. In: Ökologisches Wirtschaften, No. 3/4, pp. 18-19

Rubik, F. / Grotz, S. / Scholl, G. (1996): Ökologische Entlastungseffekte durch Produktbilanzen. Karlsruhe: Landesanstalt für Umweltschutz

Rubik, F. / Teichert, V. (1997): Ökologische Produktpolitik. Von der Beseitigung von Stoffen und Materialien zur Rückgewinnung in Kreisläufen. Stuttgart: C.E. Poeschel

Ryding, S.-O. (1998): Product Ecology Project.

Ryding, S.-O. / Steen, B. (1991): The EPS system: A PC-based system for development and application of environmental priority strategies in product design – from cradle to grave. Stockholm: IVL report B1022 - Institutet för vatten- och luftvårdsforskning

Saur, K. (1995): Kurzfassung zum Abschlußbericht "Ganzheitliche Bilanzierung eines Laugenbehälters aus Polypropylen bzw. Stahlblech". München: study on behalf of Bosch and Siemens Hausgeräte GmbH

Schafhausen, F. (1998): Der Streit um die Bewertung? Worauf können sich Industrie, Wissenschaft, gesellschaftliche Gruppen und Staat einigen? In: UTECH (ed.) (1998): Ökobilanzen VI. Berlin: FGU

Schmidt, M. (1998): Kein Platz für Produktbezug?. In: Ökologisches Wirtschaften, No. 3/4, pp. 16-18

Schmidt-Bleek, F. (1994): Wieviel Umwelt braucht der Mensch? - MIPS - Das Maß fuer oekologisch Wirtschaften. Berlin-Basel-Boston: Birkhaeuser

Schmitz, S. et al. (1995): Ökobilanz für Getränkeverpackungen. Vergleichende Untersuchung der durch Verpackungssysteme für Frischmilch und Bier hervorgerufenen Umweltbeeinflussungen. Berlin: Umweltbundesamt UBA-Texte 52/95

Scholl, G. (1999): Industry approaches to Life Cycle Assessment (LCA). In: Sikdar, S. (ed.): Tools and Methods of Pollution Prevention. Dordrecht/NL et al.: Kluwer Academic Publishers [forthcoming]

Schön, D.A. (1983): The Reflective Practitioner. How Professionals Think in Action. New York: Basic Books

SETAC (1991): A technical Framework for Life-Cycle Assessment. Vermont/USA: Workshop Report from the Smugglers Notch - August 1990

SETAC (1993): Guidelines for Life-Cycle Assessment: A "Code of Practice". Workshop Report from Sesimbra - August 1993. Brussels: SETAC

SETAC (1993): Guidelines for Life-Cycle Assessment: A "Code of Practice". Brussels: SETAC-Brochure

SETAC Europe (1997): Simplifying LCA. Just a Cut? Brussels: SETAC Europe

Simon, H.A. (1947): Administrative Behavior: A Study of Decision-Making Processes in Administrative Organization. New York: Macmillan

Simon, H.A. (1977): The New Science of Management Decisions. Englewood Cliffs, N.J.: Prentice-Hall

Simon, H.A. / Dantzig, G.B./ Hogarth, R./ Plott, C.R. / Raiffa, H. / Shelling, T.C. / Shepsle, K.A. / Thaler, R./ Tversky, A. / Winter, S. (1987): Decision Making and Problem Solving. In: Interfaces, No. 5, pp. 11-31

Singhofen, A. (1996): Introduction into a Common Format for Life-Cycle Inventory Data - Status Report. Brussels: SPOLD

Sjöberg, Å. / Ekvall, T. / Ölund, G. (1997): LCA in the Environmental Management System at Perstorp Flooring. Brussels: 5th SETAC-Europe LCA Case Studies Symposium

Smith, D. et al. (1998): Adoption by Industry of Life Cycle Approaches. Its Implications for Industry Competitiveness. London: Kogan Page

SPOLD (1995): Directory of Life Cycle Inventory Data Sources. Brussels: publication of SPOLD

Steen, B. et al. (1995): SPINE-A database structure for LCAs. Göteborg/S: IVL-REP (L95/196) B1227

Steinle, C. / Thiem, H. / Böttcher, K. (1998): Umweltschutz als Erfolgsfaktor - Mythos oder Realität? Ergebnisse einer empirischen Studie. In: Zeitschrift für Umweltpolitik und -recht, No. 1, pp. 61-78

SustainAbility / SPOLD / Business in the Environment (1993): The LCA Sourcebook. A European Business Guide to Life-Cycle Assessment. London: SustainAbility

The Nordic Council (1992): Product life cycle assessment: principles and methodology. Stockholm/Copenhagen: AKA-PRINT AS

Tillman, A.-M. (1998): Personal communication.

Tillman, A.-M. et al. (1991): Packaging and the Environment. Life-cycle analyses of selected materials. Quantification of environmental loadings. Göteborg/S: Chalmers Industriteknik

Tolbert, P.S. / Zucker, L.G. (1996): The Institutionalization of Institutional Theory. In: Clegg, S.R. / Hardy, C. / Nord, W.R. (eds.): Handbook of Organization Studies. London: Sage, pp. 175-190

Troge, A. (1997): Zur Bedeutung von Ökobilanzen. In: UBA (1997): Materialien zu Ökobilanzen und Lebensweganalysen. Berlin: UBA-Texte 26/97, pp. 3-11

UBA <Umweltbundesamt> (1988): Vergleich der Umweltauswirkungen von Polyethylen- und Papiertragetaschen. Berlin: UBA-Texte 5/88

UBA <Umweltbundesamt> (1993): Ökobilanz Rapsöl. Berlin: UBA-Texte 4/93

UBA <Umweltbundesamt> (1997a): Materialien zu Ökobilanzen und Lebensweganalysen. Berlin: UBA-Texte 26/97

UBA <Umweltbundesamt> (1997b): Nachhaltiges Deutschland. Wege zu einer dauerhaft-umweltgerechten Entwicklung. Berlin: Erich Schmidt

UBA <Umweltbundesamt> (1998): UBA-Verfahrensregeln zur Durchführung von Ökobilanzen. Berlin: manuscript of UBA

US National Researach Council (1991): Improving Engineering Design: Designing for Competitive Advantage. Washington, D.C.: National Academy Press

VCI Projektgruppe "Ökologische Bewertung" (1994): VNCI-Modell: Protocoll of the results as of March 23, 1994, NAGUS AA3/UA2

Vickers, G. (1965; 3rd ed. 1995): The Art of Judgement. London: Sage

VROM [Ministry of Housing, Physical Planning and the Environment] (ed.) (1994): Policy Document on Products and the Environment. The Hague/NL: own publication

WBCSD (1995): Achieving Eco-Efficiency in Business, Report of the World Business Council for Sustainable Development, Second Antwerp Eco-Efficiency Workshop, March 14-15, 1995, prepared by Robert U. Ayres. Conches-GE: WBCSD

Weidema, B. / Pedersen, R.L. / Krüger, I. (eds.) (1993, 2.A): Environmental Assessment of Products - A Textbook on Life Cycle Assessment. Helsinki/Finland: UETP-EEE - the Finnish Association of Graduate Engineers TEK

White, A.L. / Shapiro, K. (1993): Life cyle assessment. A second opinion. In: Environmental Sience and Technology, No. 6, pp. 1016-1017

Wolff, R. (1981): Der Prozeß des Organisierens. Zu einer Theorie des organisationalen Lernens. Spardorf: René Wilfer

Wolff, R. (1996): Von Umweltmanagement zu Umweltstrategien. Gothenborg/S: Gothenburg Research Institute - GRI Working Paper

Wrisberg, N. / Gameson, T. (eds.) (1998): CHAINET Definition Document. Final version. Leiden/Seville: Report delivered by Leiden University / IPTS

Yin, R.K. (1994): Case-study research. Design and methods. Newsburg Park CA: Stage Publications

Zoboli, R. (1998): Implications of Environmental regulation on industrial innovation: The case of end-of-life vehicles. Seville: study on behalf of European Commission DG JRC-IPTS [internal paper, forthcoming]

Abbreviations

Abbreviation	English name	Original name
ANPA	National Agency for the Protection of the Environment (I)	L'Agenzia Nazionale per la Protenzione dell' Ambiente (I)
APME	Association of Plastics Manufacturers in Europe	–
B.A.U.M.	The German Environmental Management Association (D)	Bundesdeutscher Arbeitskreis umweltbewußtes Management (D)
BDI	Federation of German Industries (D)	Bundesverband der Deutschen Industrie (D)
BLCA	Bottleneck LCA	–
BMU	Ministry for the Environment (D)	Bundesumweltministerium (D)
BS	British Standard	–
BSH	Bosch and Siemens Hausgeräte (D)	Bosch und Siemens Hausgeräte (D)
BUWAL	Federal Office for Environment, Forest and Landscape (CH)	Bundesamt für Umwelt, Wald und Landschaft (CH)
CEC	Corporate Environmental Controller	–
CFC	Clorofluorcarbon	–
CH	Switzerland	Schweiz
CIT	–	Chalmers Industriteknik (S)
CMER	Center for the Management of Environmental Resources (F)	–
CPM	Centre for Environmental Assessment of Product and Material Systems (S)	*Centrum för produktrelaterad miljöanalys*
D	Germany	Deutschland
DAPI	Department of Environment and Industrial Policies (I)	Direzione Ambiente e Politiche Industriali (I)
DG	Directorate General	–
DGXI	Directorate General for Environment, Nuclear Safety and Civil Protection	–

DIN	German Standardisation Institute	Deutsches Institut für Normung
DIS	Draft International Standard	–
EAP	Environmental Action Programme	
EBC	Ericsson Business Communication (S)	–
EC	European Community/ies or European Commission	–
ECU	European Currency Unit	–
EDIP	Environmental Design of Industrial Products	
EEB	European Environmental Bureau	–
EMAS	European Eco-Management and Audit Scheme	–
EMPA	Swiss Federal Laboratories for Materials Testing and Research (CH)	Eidgenössische Materialprüfanstalt (CH)
EMS	Environmental Management System	–
EPA	Environmental Protection Agency (USA)	–
EPS	Environmental Priority Strategies	–
EU	European Union	–
F.A.RE.	Fiat Auto Recycling	–
FAS	Fatty alcohol sulphate	–
FTU	Foundation Labour-University (B)	Fondation Travail-Université (B)
GATT	General Agreement on Tariffs and Trade	–
GDP	Gross Domestic Product	–
GNP	Gross National Product	–
GRI	Gothenburg Research Institute (S)	–
HCB	Holderbank Cement and Beton (CH)	Holderbank Zement und Beton (CH)
HSE	Health, Safety & Environment	–
I	Italy	Italia
ICC	International Chamber of Commerce	–
IDPC	Integrated Development Plan of Components	–
IFP	The Swedish Institute for Fibre and Polymer Research (S)	Institutet för Fiber- och Polymerteknologi (S)
IÖW	Ecological Economics Research Institute (D)	Institut für ökologische Wirtschaftsforschung (D)
IPP	Integrated Product Policy	–

IPTS	Institute for Prospective Technological Studies	–
ISO	International Organization for Standardisation	–
IVL	Swedish Research Institute for the Environment	Institutet för Vatten- och Luftvårdsforskning (S)
IWÖ	Institut for Economy and Ecology (CH)	Institut für Wirtschaft und Ökologie (CH)
KJS	Kraft Jacobs Suchard	-
KrWAbfG	Waste Management and Product Recycling Act	Kreislaufwirtschafts- und Abfallgesetz
LAS	Linear alkylbenzene sulfonate	–
LCA	Life Cycle Assessment	–
LCC	Life Cycle Costing	–
LCI	Life Cycle Inventory	–
LCT	Life Cycle Thinking	–
LGUT	Landis & Gyr Utilities (CH)	Landis & Gyr Europe, Segment Utilities (CH)
MIPS	Material intensity per service unit	Materialintensität pro Serviceeinheit
NACE	–	Nomenclature des Activités économiques dans les Communautés Européennes
NGO	Non-Governmental Organisation	–
NUTEK	Swedish National Board for Industrial and Technical Development (S)	Närings- och teknikutvecklingsverket (S)
ÖBU	Swiss Association for Environmentally Conscious Management	Schweizerische Vereinigung für ökologisch bewußte Unternehmensführung (CH)
OECD	Organisation for Economic Cooperation and Development	–
PE	Polyethylene	–
PEMS	Product Environmental Management System	–
PEP	Product-Eco-Performance	–
PET	Polyethylene terephthalate	–
PP	Polypropylene	–
PPP	Polluter Pays Principle	–
PVC	Polyvinylchloride	–
R&D	Research & Development	–
R, D & D	Research, Development and Design	-

RAL	German Institute for Quality Assurance and Labelling	Deutsches Institut für Gütesicherung und Kennzeichnung
S	Sweden	–
SAE	Society of Automotive Engineers	–
SBU	Strategic business units	–
SETAC	Society for Environmental Toxicology and Chemistry	–
SLCA	Streamlined Life Cycle Assessment	–
SME	Small and Medium Sized Enterprise	–
SOP	Standard Operating Procedure	–
SPOLD	Society for the Promotion of LCA Development	–
SPRU	Science Policy Research Unit (UK)	–
TMP	Technical Environmental Planning Department	*Avdelningen för Teknisk Miljöplanering*
UBA	Federal Environmental Agency (D)	Umweltbundesamt (D)
UK	United Kingdom	–
VAT	Value Added Tax	–
VCI	Association of Chemical Industry (D)	Verband der Chemischen Industrie (D)
VOC	Volatile Organic Compounds	–
W & CA	Washing and Cleaning Agents	–
WBCSD	World Business Council for Sustainable Development	–

Printing: Druckhaus Beltz, Hemsbach
Binding: Buchbinderei Schäffer, Grünstadt